IMPACT OF IMPROVED OPERATION AND MANAGEMENT ON COHESIVE SEDIMENT TRANSPORT IN GEZIRA SCHEME, SUDAN

Ishraga Sir Elkhatim Osman

Thesis committee

Promotor

Prof. Dr E. Schultz
Emeritus Professor of Land and Water Development
UNESCO-IHE Institute for Water Education
Delft, the Netherlands

Co-promotors

Prof. Dr Akode Osman
Professor of Hydraulics and Water Resources
Civil Engineering Department
University of Khartoum
Khartoum, Sudan

Dr F.X. Suryadi
Senior lecturer in Land and Water Development
UNESCO-IHE Institute for Water Education
Delft, the Netherlands

Other members

Prof. Dr R. Uijlenhoet, Wageningen University
Prof. Dr N.C. van de Giesen, Delft University of Technology
Prof. Dr Abdin Salih, University of Khartoum, Sudan
Dr Nestor Mendez, UCLA, Barquisimeto, Edo. Lara. Venezuela

This research was conducted under the auspices of the SENSE Research School for Socio-Economic and Natural Sciences of the Environment.

IMPACT OF IMPROVED OPERATION AND MAINTENANACE ON COHESIVE SEDIMENT TRANSPORT IN GEZIRA SCHEME, SUDAN

Thesis

submitted in fulfilment of the requirements of
the Academic Board of Wageningen University and
the Academic Board of the UNESCO-IHE Institute for Water Education
for the degree of doctor
to be defended in public
on 3 September 2015 at 11 a.m.
in Delft, the Netherlands

by

Ishraga Sir Elkhatim Osman
Born in Port Sudan, Sudan

CRC Press/Balkema is an imprint of the Taylor & Francis Group, an informa business

© 2015, Ishraga Sir Elkhatim Osman

Published by
CRC Press/Balkema
PO Box 11320, 2301 EH Leiden, The Netherlands
e-mail: Pub.NL@taylorandfrancis.com
www.crcpress.com - www.taylorandfrancis.com

ISBN 978-1-138-02880-7 (Taylor & Francis Group)
ISBN 978-94-6257-434-2 (Wageningen University)

Table of contents

Acknowledgements

I owe an incalculable debt to my promoter Prof. em. E. Bart Schultz for his endless support, constructive comments, enthusiastic guidance and continuous encouragement that are beyond the call of duty throughout the course of this work. I also would like to thank Dr. F.X. Suryadi as co-supervisor for his fruitful discussions and support during the period of the research. Special thanks to Prof. Akode Osman, the local supervisor, for his time in discussion, great contribution and for his valuable advice during this work.

The study was carried out as part of a project titled "In search of sustainable river basins and basin-wide solidarities in the Blue Nile River Basin". I would like to express my deep and sincere gratitude to all project coordinators and all PhD students in the project. Special thanks to Prof. Pieter Van der Zaag, Prof. Stefen Uhlenbrook and Dr. Belay for their advice and feedback during my study. Many thanks go to Herman for having nice discussions and sharing data on the study area.

I am grateful to Dr. Yasir Mohamed for the valuable information regarding the operation of Gezira Scheme and for his support during the fieldwork. I also would like to thank the support of the Hydraulic Research centre staff, Abo Obida and Elterifi during the fieldwork period. Special thanks to Dawood and his family for their support in data collection and nice hospitality in Toman Minor Canal during fieldwork. Many thanks also to Senad, Abdel Karim and Abdel Azim, the MSc Students, the University of Gezira for their help during the fieldwork.

I am greatly indebted to Dr Alessandra Crosato, Chair Group of Hydraulic Engineering and River Basin Development at UNESCO-IHE for offering expert guidance and suggestions regarding the model. My thanks extend to Prof. Ahmed Salih, Prof. Abas Abdalla, Prof. Hassan Fadol and Dr. Adil Khider for their useful comments regarding the sediment and the operation of the scheme. I am grateful to the University of Khartoum staff and to Prof. Abdin Salih and Prof. Gamal Abdo for their useful advices.

The study would not have been possible without the financial support from the Netherlands Organization for International Cooperation in Higher Education (NUFFIC) and the support of the International Foundation for Science (IFS) for the fieldwork. I extend my thanks to Jolanda Boots and staff members of the UNESCO-IHE Institute for Water Education for the help and logistic support that they provided for this study.

I enjoyed the company of my friends Reem, Eman Fedol and Eman Elshikh. It would have been much more difficult to live away from my family without their company.

I am thankful to my mother Mariam and my brother Ashraf for their moral support and encouragement and for their belief in me. I am thankful to the cooperation, continuous support and endless patience of my husband, Walaa and for my daughters during this study period. Finally, this thesis is dedicated to my father, who passed away last year, he always gave me the motivation to do the best in this research and taught me the sense of satisfaction.

Acknowledgements

Summary

Efficient operation and maintenance of irrigation schemes are needed for improving the hydraulic performance of the canals, enhancing the crop yields and insuring sustainable production. There is a great need to enhance the researches and for a variety of tools such as water control and regulation equipment, decision support systems, as well as field surveys and valuation techniques. Water management becomes difficult when dealing with sediment transport in irrigation canals. Most of the studies simulate the sediment transport of relatively coarse grain sizes. The sediment problem in irrigation canals becomes more complicated when dealing with cohesive sediment transport. Therefore, more research is needed to enhance the understanding of the behaviour of cohesive sediment transport under a variety of operation conditions.

This study has been carried out in the Gezira Scheme in Sudan. The scheme, which is one of the largest irrigation schemes in the world under a single management, is located in the arid and semi-arid region. The scheme is chosen as a case study since it can act as a model for similar irrigation schemes. The scheme has a total area of 880,000 ha and uses 35% of Sudan's current allocation of Nile waters. This represents 6 – 7 billion m^3 per year. The scheme is irrigated from the Blue Nile River, which is characterized by its high load of fine sediment. The scheme is facing severe sediment accumulation in its irrigation canals, which represents a challenge to those responsible for the operation and maintenance of the canals. Each year large investments are required to maintain and to upgrade the canal system to keep it in an acceptable condition.

A large quantity of cohesive sediment enters the scheme every year. According to previous studies, about 60% of the sediment deposits in the irrigation canals. The sediment accumulation in the canals reduces the canal conveyance capacity, causes irrigation difficulties, creates inequity and inadequate water supply and increases the rate of aquatic weed growth. The sedimentation problems are not only seriously affecting the performance of the irrigation canals, but are also jeopardizing their sustainability, as well as affecting crop production. Two canals in the scheme have been selected to be studied in detail: Zananda Major Canal, which takes water from Gezira Main Canal at 57 km from the offtake at Sennar Dam, and Toman Minor Canal at 12.5 km from the offtake of Zananda Major Canal.

The hypothesis of the study postulates that the operation and maintenance of an irrigation scheme has a major influence on the hydrodynamic behaviour of canals and hence on sediment movement and deposition. The aim of this study was to improve the operation and maintenance procedures for better sediment and water management. This can be achieved through better understanding of the sediment processes in the irrigation canals of the Gezira Scheme and to understand clearly the link between irrigation system operation and resulting system performance in terms of transport of cohesive sediment.

Data collection and field measurements have been conducted during the flood season between June and October in 2011 and 2012. Sediment sampling and water level measurements have been conducted on a daily basis at selected locations. The manually recorded water levels include about 1080 readings per year. In addition about 1290 sediment samples were analysed for different locations during the study period. Cross-sectional surveys have been performed at the beginning and end of the flood season to address the spatial and temporal variation of the sediment deposition in the canals under study and to detect changes in the bed profile. The head regulator and outlet control structures were calibrated by using the measured stage-discharge relationships. More elaboration is given to the properties of cohesive sediment and identification of the dominant factors that cause deposition in irrigation canals. Sediment properties were tested such as grain size distribution, mechanical and physico-chemical properties of the sediment. The irrigation schedules, cropped area and sowing dates for different crops were reported. Other data such as canal design data, historical data of the sediment and flow for certain canals were reviewed.

The analysis of the data indicates a variation of the water level along the canals under study. It should be noted that the operation control in Gezira Scheme is by using upstream control structures. The field data show that the flow release in the system is not regularly adjusted in a systematic way to meet the demand and maintain the required water level. Continuous change in gate setting results in instability of the water level. This situation became worse with more sediment deposition. The water level has been raised far above the design level and there is lapse in working levels especially at the major and minor canals. The rise is found to be about 1.6 and 1.2 m above the design level at the head of the major and minor canals under study. Furthermore, reduction in the water depth has been detected along the canals as result of bed rise and enlarging of canal sections due to improper desilting. The results demonstrate that the supply of water was extremely large during the flood season of 2011 compared to the actual crop water requirement, especially during the period of high sediment concentration. The delivery performance ratio indicated an oversupply at the major canal in 2011 during most of the time. The study also provides some valuable insight into the nature of sediment in Gezira Scheme.

There is a limitation in the existing models that deal with fine sediment transport in irrigation canals. Most of the sediment transport models are developed for estuaries and rivers. Therefore there was a great need to develop a simple but effective numerical model that incorporates control structures to simulate the fine sediment transport in irrigation canals. Although there are similarities between rivers and irrigation canals, irrigation canals are different. The presence of a large number of flow control structures and the high influence of the side banks on the velocity distribution create some differences in both types of channels. Hence, it was important to develop a model dealing with fine sediment in irrigation canals, including different types of hydraulic structures.

In line with this the one dimensional numerical model Fine SEDiment Transport (FSEDT) dealing with fine sediment transport in irrigation canals has been developed. The model has been used as a tool to study the mechanism of water and sediment flow under different operation and maintenance scenarios. The water surface profile has been predicted by using the predictor corrector method to solve the gradually varied flow equation. The prediction of sediment concentration is based on the solution of the one dimensional advection-diffusion equation. The bed material exchange was determined based on the Partheniades (1962) and Krone (1965) equations. The change in bed level was computed based on the sediment mass balance equation that was solved numerically by using the finite difference method. The model has been applied in the Gezira Scheme. On the basis of the field data the model has been calibrated and validated. The predicted bed profiles depict good agreement with the measured ones. The model is capable to predict the bed profile for any period of simulation. The model can predict the sediment concentration hydrograph at different points within a canal reach, in addition to the total volume of the sediment deposition in the reach. The output of the model can be presented in tabular or graphical form.

The sediment transport in the irrigation canals has been simulated by adopting different scenarios. The interrelationship between water flow and sediment transport in the irrigation canals under changing flow conditions has been investigated. Two scenarios of operation were tested at the major canal under study. The model evaluated the indent system that has been applied in Gezira Scheme for many years in regard to sediment deposition. Another proposed scenario based on crop water requirement was also tested. In addition, operation under future changed conditions in case of reduction in the sediment concentration was tested. The different operation scenarios have been compared with the existing condition based on data collected during the flood season in 2011 in terms of sedimentation. Based on this, the following remarks are made:

- the effect of varying crest settings of the movable weirs has been investigated and less sediment deposition was found to occur when the crest level was set at its lowest position. The sediment transport in the canals is influenced by the operation of the hydraulic structures, especially upstream of movable weirs. The effect is extended to about 3 km upstream of the weir;

- for many years the indent system of water allocation was applied in the Gezira Scheme based on duty and cropped area. However, this system of operation has been absent during the last years. The simulation of the suspended sediment transport reveals that less sediment will be deposited if this system is applied. The reduction in sediment deposition was found to be 34 and 40% in the first and second reaches respectively when related to the actual situation in 2011. More deposition took place when the model was set with the design canal profile. The slope of Zananda Major Canal became 13 cm/km and 18 cm/km for the first and second reaches respectively, while the design slopes were 10 cm/km and 5 cm/km for the first and second reaches respectively. Steepening of the slope especially in the second reach is an indicator of improper desilting campaigns;

- the reduction of the water delivery during the period of high concentration between 10 July and 10 August, based on the crop water requirement results in reduction in the sediment deposition by 51 and 55% for the first and second reaches respectively when compared to the situation in 2011;

- the reduction of the Blue Nile River sediment concentration by 50% as result of the construction of the Ethiopia Renaissance Dam and/or improvement in the land use has been simulated. The results of the simulation of the suspended sediment transport at the major canal indicate that the deposition will be 74 and 81% lower for the first and second reaches respectively when compared with the situation in 2011.

At the minor canals, the night storage weirs were designed as cross structures. The idea behind the night storage system was to store water during the night by closing all field outlet pipes and the gates of the cross structures along the minor canal at 6:00 pm and releasing them at 6:00 am. Although this system has been vanished to keep pace with crop intensification and to cope with the deterioration of the water supply due to the poor maintenance of the canals, this scenario has also been simulated. The hydrodynamic flow in the canals during the filling time has been simulated by using the DUFLOW model since the model can be applied for unsteady flow. A spreadsheet has been designed to predict the deposition every hour based on the output of the DUFLOW model. The night storage system has been compared with the continuous system regarding the sediment transport in addition to other scenarios. It was found that:

- the continuous system reduces the amount of deposited sediment by 55% compared to the night storage system;

- about 29% of the sediment was reduced in 2011 when the system was operated based on crop water requirement;

- the deposition slightly increased with reduced capacity of the field outlet pipes.

The increase of sedimentation is 22% and 31% when the capacity of the field outlet pipes is reduced by 75% and 50% respectively.

The main findings and the contributions that are made by this study:

- the study comes up with a model dealing with cohesive sediment in irrigation canals for effective sediment and water management, which can be applied widely for similar irrigation schemes dealing with fine sediment;
- it is possible to improve the sediment and water management by improving the operation and maintenance. The high irrigation efficiency is tending to mitigate the inflow sediment load and as a consequence less deposition is expected;
- the study comes up with strategies of water management that can reduce the deposition in irrigation canals by operating the system continuously based on crop water requirement at the period of high sediment concentration with the field outlet pipes operating at their full capacity.

The absence of proper maintenance activities and water management has a prominent role in increasing the deposition along the irrigation canals in Gezira Scheme. Improving the operation and maintenance is not the only way to mitigate the sedimentation in the irrigation canals. A great consideration needs to be given to improve the design since conditions based on the original design have been changed with time such as the operation system (night storage system, indent system), cropping intensity and geometry of the canals. In other words, rehabilitation of the system will not be one of the solutions to mitigate the accumulation of the deposition along the canals but the system itself needs remodelling. The developed model can be used to assess the new design and to evaluate the proposed management plans in terms of transport of cohesive sediment.

1 Introduction

1.1 General

Improving the water management of irrigation schemes via sediment management is needed for adequate water supply and food production. The efficiency of water allocation and application can be improved through diagnosis of water and sediment management at the irrigation district level.

There are numerous studies dealing with sediment management in irrigation canals. Jinchi *et al.* (1993) in his study found that sediment degradation and aggregation processes in irrigation canals greatly depend upon the hydrograph of water and sediment discharge. With some adjustments of the processes in certain time intervals, it is possible to transport as much as possible sediments into a further area for deposition or for farmland usage. Bhutta *et al.* (1996) investigated clearance activities in Pakistan and found that if the desilting campaign will be done in the upper two-thirds of the canal, it would greatly improve hydraulic performance of the canals. Belaud and Baume (2002) developed a methodology based on numerical modelling and illustrated it for a secondary network in Sangro Distributaries System in South Pakistan. Improvements in the design and desilting process were proposed in order to preserve the equity longer. Depeweg and Paudel (2003) evaluated the design of Sunsari Morang Irrigation System in Nepal for different operation and maintenance plans and their effectiveness on sediment transport. Sherpa (2005) applied the SETRIC model to simulate sediment transport in irrigation canals in Nepal, while Sutama (2010) applied the same model in an irrigation scheme in Indonesia. Both studies addressed the applicability and versatility of the model for different conditions of operation and sediment input in irrigation canals. Jian (2008) developed a mathematical model, applied the model to simulate the sediment in irrigation canals and found that it can be used to predict the non-uniform sediment movement in irrigation canals. Paudel (2010) found that it is possible to reduce the sediment deposition problem by proper design and management of the system. Munir (2011) suggested an improvement in the canal operation in his study on the USC-PHLC Irrigation System in Pakistan. He found that the sediment deposits during low crop water requirement periods can be re-entrained during peak water requirement periods. Most of these studies were dealing with non-cohesive sediment. There is still limitation in the research that deals with fine sediment transport in irrigation canals.

The dynamics of cohesive sediment transport are mainly treated empirically since they are affected by numerous parameters that cannot be determined theoretically. Moreover, the combination of hydrodynamic, cohesive sediment properties and biological processes make the prediction of cohesive sediment dynamics difficult. The use of modelling tools improves the understanding of fine sediment dynamics. Therefore, a model to simulate the suspended sediment transport cannot be developed without deep understanding to the properties of the sediment under the study. This understanding is most expediently developed through the field data analysis and formulation of a conceptual model.

This study has been carried out in Gezira Scheme, Sudan. The scheme is facing severe sedimentation problems in its irrigation canals which lead to unreliable water supply. Winterwerp and Kesteren (2004) defined sedimentation as the net increase in bed level and the sedimentation rate is the deposition rate minus the erosion rate. The operation and maintenance becomes challenging in the scheme. The aim of this study was to improve the operation and maintenance through better sediment and water management. To achieve this a one dimensional model has been developed based on sub-critical, quasi-steady flow that can simulate sediment transport under non-equilibrium conditions. A great effort has been made for adequate data collection and field measurements in order to investigate the behaviour of the sediment in the irrigation canals. The model was carefully developed to simulate the sediment transport based on different options of operation.

1.2   Scope of the study

The sedimentation problems and their effect on the operation are highlighted in this study. The consequences of the sediment accumulation along the canals on the stability of the water level have been described. The difference between cohesive and non-cohesive sediment transport has been explained. Great attention is given to the properties of the cohesive sediment and to the hydrodynamics of the suspended sediment transport. The one dimensional numerical model that has been developed to simulate the suspended sediment transport of fine material has been calibrated and validated with field data that were collected in 2011 and 2012. The sensitivity of different sediment parameters on the deposition has been investigated. The deposition process has been investigated by using the modelling approach. The mechanisms of water flow and sediment transport have been investigated under different operation scenarios. The evaluation of the scenarios resulted in different options for sediment and water management.

1.3   Structure of the thesis

Chapter 1 gives an overview of the research, the scope of the study and the structure of the thesis.

Chapter 2 provides background information of the Blue Nile River Basin and the irrigation schemes along it in Sudan, the rational of the study, research questions, hypothesis and objectives.

Chapter 3 provides general information on the Gezira Scheme, the sedimentation problems in the scheme, water and sediment management and reviews of the previous studies on the Gezira Scheme.

Chapter 4 presents a summary of the most important processes of the cohesive sediment, sediment transport under non-equilibrium conditions and gives an overview of cohesive sediment transport models.

Chapter 5 presents the data collection and field measurements, the techniques used in the measurements, laboratory works and detailed analysis of the data collected and scheme analysis.

Chapter 6 presents the conceptual approach that has been applied in the development of the model, the principle followed in the hydrodynamics, sediment transport and morphological change computation to predict the fine sediment transport.

Chapter 7 describes the required considerations in the model setup, calibration, validation and verification. It addresses the sensitivities of the different parameters to the deposition, the dynamics of suspended sediment transport and quantifies the deposition in the studied canals by applying different scenarios.

Chapter 8 evaluates the findings of this study, gives options of sediment and water management and proposes a way forward.

2   Background and objectives

2.1   Background information on the Blue Nile River Basin

The Nile River is the longest river in the world; the river originates from two distinct geographic zones (Equatorial lakes plateau and Ethiopia high plateau). The main branches of the Nile River are the White Nile River, the Blue Nile River and the Atbra River as presented in Figure 2.1. The Blue Nile provides the greater part, about 60% of the flow of the main Nile River with a mean annual discharge of 50 billion m³ (Elfaki, 2008). The Blue Nile originates from the Ethiopia high plateau at an elevations of 2000-3000 m+MSL (mean sea level) with several peaks up to 4000 m+MSL or more (Sutcliffe and Parks, 1999). The high rainfall over the Ethiopia highlands in a single season and the steep topography give rise to a relatively high and concentrated runoff. The plateau drops steeply to the Sudan plains where there are many isolated outlaying hills and the vegetation cover is relatively sparse because of the short rainy season.

Figure 2.1. Nile River Basin (Hagos, 2005)

Figure 2.2 demonstrates that the Blue Nile River transports a huge amount of sediment compared to the other rivers originating from the Ethiopia highlands. The upstream of the Blue Nile River Basin witnesses severe erosion and loss of top soils, with sedimentation and morphological changes in the downstream of the basin. These are attributed to the degradation in the upper basin that leads to high velocity of the flow and increases the flood hazard, as well as the sediment load downstream. Figure 2.3 illustrates the problem analysis upstream and downstream of the Blue Nile River Basin. About 140 million tons of the sediment are annually transported by the Blue Nile River to Sudan (Hagos, 2005). This high sediment load has a major influence on the design and operation of the reservoirs built across the river (Sennar and Roseires dams) and the irrigation schemes. Sennar and Roseires reservoirs have lost 60% and 34% of their storage capacities of 0.93 and 3.02 BCM (billion cubic metres) respectively due to sediment deposition (Gismalla, 2009). Consequently, this reduction in the storage capacity has affected the water supply to the irrigation schemes, which was the main purpose of the construction of these dams besides hydropower generation.

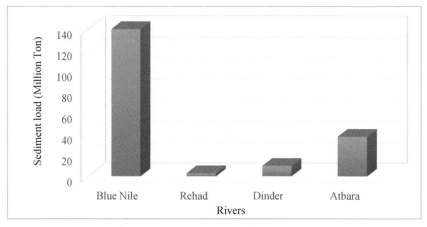

Figure 2.2. Total annual sediment load (million tons) for different rivers

The problem of sediment deposition in irrigation canals represents a challenge to those responsible for the operation and maintenance of these canals in Sudan. This research focuses on the impact of the sediment downstream of the Blue Nile River Basin and how it affects the irrigation system as is shown in Figure 2.3.

This study is a part of the 'In Search of Sustainable Catchments and Basin-wide Solidarities; Trans-boundary Water Management of the Blue Nile River Basin' project. The aim of this project is to generate a better understanding of the agronomic, hydrological, environmental and socio-economic impacts of the improved land management along the Blue Nile River. It focuses on the impact of land use management upstream (Ethiopia) on sedimentation rates downstream (Sudan). In this study, the Gezira Scheme has been chosen as a sink of eroded sediment from upstream of the Blue Nile. The scheme acts as a model for other large and small scale schemes and its significant role in the economy and socio economy of Sudan. Operation and maintenance procedures in Gezira Scheme have been investigated in order to find an appropriate strategy to reduce the sediment accumulation in the canal systems. Through links with other researchers this research leads to better understanding of the downstream impact of the improvement of land use management. The research contributes to achieve food security and poverty eradication of local communities downstream, which is the overall development objective of the project.

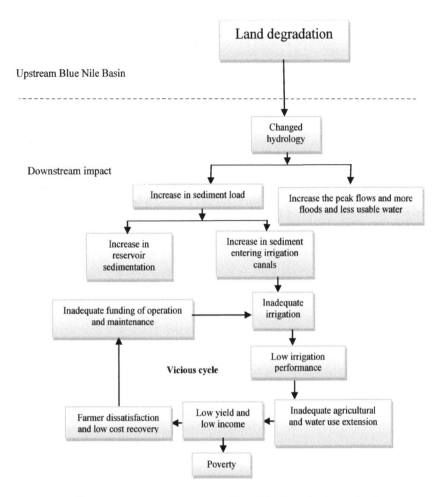

Figure 2.3. Effect of erosion upstream of the Blue Nile River Basin on the downstream

A number of studies have addressed the sedimentation of the Blue Nile River. Hussin (2006) studied the sedimentation in the Blue Nile River and found that most of the sediment in cohesive grain size is clay and silt, transported as suspended sediment with higher concentration than the sand concentration as illustrated in Figure 2.4. The wash load in the graph is referred to clay and silt. Steenhuis *et al.* (2009) developed a runoff and sediment loss model for prediction of the sediment concentration. This model depends on the assumption that sediment is carried by overland flow. It is recognized that the model captures the high sediment concentration on the rising limb and the lower concentration on the falling limb of the flow hydrograph. The decrease in sediment concentration after reaching the peak occurs before the peak of the discharge. That is related to time when the river basin becomes fully wetted up and becomes covered with vegetation and the erosion reduces. Billi and Ali (2010) analysed the pattern of suspended sediment concentration, discharge and grain-size variation with flow based on field measurements at the early 1960s. The suspended sediment yield was calculated and compared with data from different sources. Billi and Ali (2010) developed a good relation between flow velocity and bed load transport rate. Omer (2011) studied the sedimentation in Roseries Reservoir and predicted the vertical and horizontal sorting of the sand and silt of sediment over time by using the Deft3D model. The effect of heightening on the trap efficiency of the reservoir was also highlighted in the study. Ali (2014) scrutinized the sedimentation along the Blue Nile River and managed to distinguish the source of the sediment from the silt that deposits in Roseries Reservoir according to the findings of Omer (2011).

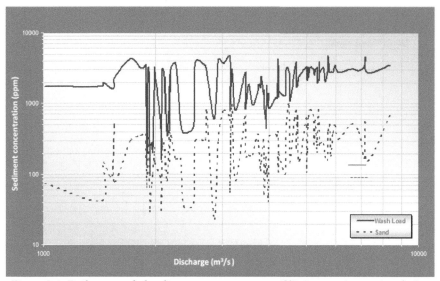

Figure 2.4. Daily suspended sediment measurement at Al Deim gauging station during the flood of 1993 (Hussin, 2006)

2.2   Overview of irrigation along the Blue Nile in Sudan

Sudan has a large irrigated agriculture sector, totalling more than 2 million ha out of about 84 million ha that is potentially arable agricultural area. The potential arable agricultural lands are used to a limited extent because of poor or absence water resources management practices. About 93% of the irrigated area concern governmental schemes; the remaining 7% belongs to the private sector. The Nile and its tributaries are the main source of water for 93% of irrigated agriculture, and of this the Blue Nile River accounts for about 67% (Central Intelligence Agency (CIA), 2004). Gravity flow is the main form of irrigation.

Large scale irrigation in Sudan was established at the beginning of the 20th century. Figure 2.5 displays the irrigation schemes along the Blue Nile River. Gezira Scheme was officially inaugurated in 1925 after the construction of Sennar Dam, starting with an area of 350,000 ha that was extended gradually to more than 880,000 ha after the development of Managil Extension in 1960s (Ahmed *et al.*, 2006). Rahad Irrigation Scheme with 63,000 ha was implemented east of the Rahad River, a seasonal tributary of the Blue Nile River. The sole source of the irrigation water in the dry season is the Blue Nile River.

Since the 1950s, the government has constructed a number of large pump schemes, most of them on the Blue Nile River. These include the Junayd Scheme on the right bank of the Blue Nile River east of the Gezira Scheme with a total area of 36,000 ha. The scheme produced only cotton until 1960, when about 8,400 ha were converted to sugarcane. In the early 1970s Al Suki Scheme was established upstream of Sennar Dam to grow cotton, sorghum and oilseeds with a total area of 36,000 ha. In the mid-1970s, the government constructed a second scheme near Sennar of about 20,000 ha. In addition to cotton and other crops such as peanuts, about 8,400 ha of the area were devoted to raise sugarcane. Several smaller Blue Nile pump irrigated schemes were established. They added more than 80,000 ha to Sudan's overall irrigated area (Central Intelligence Agency (CIA), 2004).

Ararso and Schultz (2008) conducted an investigation on Sub-Saharan Africa taking six sample countries (Cameroon, Democratic Republic of Congo, Ethiopia, Nigeria, South Africa and Sudan). The result of the analysis depicted that Sudan has the lowest cropping intensity among others. The country is suffering from food shortage and might also continue to suffer in future. They also concluded that the degree of development of water resources is low. The availability of land by itself is not sufficient to achieve food security unless the productivity of the land resources is increased through more effective water management measures. Therefore, there is a gap between the potential to increase food production and the problem of food insecurity for the growing population in the country.

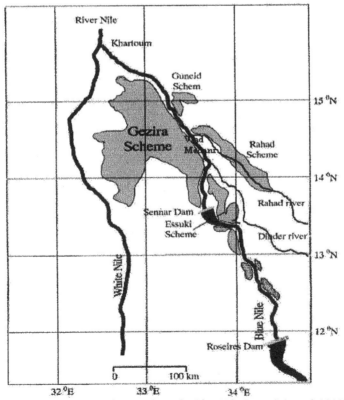

Figure 2.5. Irrigation schemes along the Blue Nile River (Mamad, 2010)

2.3   Rational of the study

The high sediment load carried by water coming from erosion in the upper basin of the Blue Nile River entering the Gezira irrigation network induces a severe problem of sediment deposition in the canals. The amount of sediment entering the network is increasing every year and the sediment problem became significant. The average annual amount of sediment removal is about 16.5 Mm^3 (million cubic metres) costing more than US$ 10 million and it is increasing in recent years according to Gesmalla (2009). More than 60% of the operation and maintenance costs go to sediment removal. Sediment deposition in front of canal offtakes and along the canals poses serious threats to the network system such as blocking the offtake pipes and gates, raising canal beds, increasing the field levels, reducing canal conveyance capacity and favours aquatic weed growth. The growth of aquatic weeds aggravates the sedimentation rate. The canal sedimentation is the cause of problematic control and operation. That results in higher water levels than the design values and leads to frequent breaking or overtopping of canal banks in addition to inadequate water supply and inequity. This is reflected in the variation in crop yield. Therefore, the productivity of the scheme reduces.

Literature dealing with sedimentation problems in irrigation canals is very limited especially when dealing with cohesive sediment. The physical processes of the cohesive sediment transport are very complicated and not clear (Lopes *et al.*, 2006). In general there is knowledge gap and more research is needed to understand the behaviour of cohesive sediment under a variety of operation conditions. The effect of the different operation on the cohesive sediment processes in irrigation canals is not well understood. Moreover, most of the models that were developed to simulate the fine sediment transport are developed for rivers or estuaries. There is an important need of a model for fine sedimentation in irrigation canals.

At the scheme level, many practical approaches have been implemented to mitigate the sedimentation problem with little success and many failures. Excavators for sediment and aquatic weeds removal (with specialized weed-cutting) are used, but there is lack of funds for maintenance to keep them working properly. Nothing has been done to improve the operation for less sediment deposition.

2.4   Research questions

To address the sedimentation problem of cohesive sediments in the Gezira irrigation canals, the following research questions were raised:
- What are the dominant factors that increase the deposition rate of cohesive sediment in the irrigation canals?
- What is the effect of different operation and maintenance methods on sediment transport in Gezira Scheme, and what is an optimum operation scenario to reduce the canal sedimentation?
- How could the operation and maintenance in the Gezira Scheme be improved to reduce the sediment deposition and provide adequate crop water supply?

2.5   Research hypothesis

The concept of the study postulates that the operation and maintenance of an irrigation scheme has a major influence on the hydrodynamic behaviour of canals and hence on the movement and deposition of cohesive sediment. It is possible to increase the productivity of the water used in Gezira Scheme through less sedimentation and hence better operation and maintenance.

2.6   Objectives of the study

The main objective is to reduce the impacts of cohesive sediment deposition in the irrigation canals of the Gezira Scheme by improving the operation and maintenance procedures.

Specific objectives

The specific objectives are:

- to understand clearly the link between irrigation system operation and the resulting system performance in terms of transport of cohesive sediment and to develop a plan for implementing needed changes;
- to understand the sedimentation processes in the irrigation canals;
- to develop a numerical model to simulate the suspended sediment transport in irrigation canals and use it as a tool to study the mechanism of water and sediment flow under different operation and maintenance scenarios, and to come up with a strategy in order to reduce the deposition of cohesive sediments in the irrigation canals in Gezira Scheme;
- to improve the reliability of the irrigation water delivery system in the Gezira Scheme considering the sedimentation problem.

3 Gezira Scheme

3.1 Background information on Gezira Scheme

The Gezira Scheme is one of the largest irrigation schemes in the world under a single management. The scheme is located between latitudes 13°30′N and 15°30′N and longitudes 32°15′E and 33°45′E, as shown in Figure 2.5. It is located between the Blue Nile and the White Nile south of Khartoum. The scheme uses up to 35% of the Sudan's current allocation of Nile water, 6.0 - 7.0 BCM per year (Ahmed *et al.*, 2008). Two main canals (Gezira and Managil canals) supply the water requirement to the scheme from Sennar Dam with a combined capacity of 354 m³/s. Gezira Scheme produces 65% of the country's cotton, 70% of wheat, 32% of sorghum, 15% of groundnut and 20% of vegetables (Elhassan and Ahmed, 2008). In addition, it contains more than 1.7 million of animal heads (these include cattle, camels, sheep and goats). This is making it one of the most important schemes for food security in the country. The total number of farmers in the scheme is about 114,000 (owned farms), among them are 12,000 female farmers (Elhassan and Ahmed, 2008).

3.1.1 Climate

Gezira Scheme is located in a semi-arid area characterized by low average annual precipitation and high evaporation. The average annual rainfall ranges from 470 mm per annum in Sennar to approximately 160 mm per annum at Khartoum, almost from June to September (Mamad, 2010). The relative humidity fluctuates from 20 to 70% and temperature varies from 5 °C in December to over 46 °C in April, with an annual mean of 28 °C. The reference evapotranspiration, ETo (Penman), at Wad Medani varies from 5.5 mm/day in December to 9 mm/day in June, with an annual average of 2,630 mm (Plusquellec, 1999). Tables 3.1 and 3.2 summarize the meteorological data for Wad Madani station, which is the nearest station to the scheme.

Table 3.1. Meteorological data for Wad Madani station in 2011

Month	Mean temperature (°C)		Relative humidity (%)	Rainfall (mm)	Sunshine (hr)	Wind speed (knots)
	max	min				
January	32.7	12.5	31	0	10.7	5
February	37.5	17.3	25	0	11.1	5
March	38.0	19.4	26	0	10.5	6
April	42.2	22.4	27	0	10.1	5
May	42.4	25.0	30	5.6	8.7	6
June	41.9	26.9	29	0.1	7.7	6
July	39.8	25.0	39	14.9	5.8	7
August	36.2	22.7	52	160	6.1	6
September	38.1	23.3	45	0.3	9.0	5
October	40.0	22.3	39	12	10.2	4
November	36.1	15.5	33	0	11.2	5
December	35.4	15.2	30	0	10.7	4

(Source: Meteorological Corporation records, 2011)

Table 3.2. Meteorological data for Wad Madani station in 2012

Month	Mean temperature (℃)		Relative humidity (%)	Rainfall (mm)	Sunshine (hr)	Wind speed (knots)
	Max	Min				
January	33.9	15.3	19	0	10.5	5
February	38.4	19.1	23	0	10.1	6
March	39.1	19.8	21	0	9.9	6
April	42.2	21.5	25	0	10.5	5
May	42.7	25.4	28	11.3	7.7	6
June	40.0	25.4	40	12.4	6.3	8
July	35.8	23.1	56	89.1	5.7	8
August	34.0	22.2	61	131	5.3	7
September	37.5	22.9	49	3.7	3.9	5
October	38.8	22.1	44	39.6	9.9	4
November	37.1	18.7	38	0.1	10.3	4
December	34.8	15.5	28	0	10.7	4

(Source: Meteorological Corporation records, 2012)

3.1.2 Topography and soil

The topography and soil of Gezira Scheme can be described as fertile flat central clay plains with a gentle slope to the North and West and are characterized by dark and heavy soils. The soil has a smectite clay fraction and is fine-textured, with 60 to 80% clay content. Smectite is the predominant mineral responsible for its swelling and shrinkage during wetting and drying. The very low water loss from this soil is due to negligible deep percolation in the field and seepage from the canals; therefore there is no need for expensive canal lining.

3.1.3 Irrigation system

The scheme is irrigated by two main canals as mentioned before: Gezira Main Canal (design capacity of 168 m³/s) and Managil Main Canal (design capacity of 186 m³/s), meeting together at K57 (57 km from Sennar Dam). At this junction, the irrigation water flow is measured by using the headworks, which distribute water to all parts of the scheme. From this point the sediment is also diverted to the scheme.

The irrigation network in Gezira Scheme consists of main and branch canals that supply water to the major canals. Then the major canals supply water to the minor canals. The water deliveries from the minor canals go to Abu Ishreen through field outlet pipes (FOP). The flow is controlled by sluice gates. The command area of each minor canal is divided into groups of fields called Nimras, arranged parallel to each other and irrigated by field canals called Abu Ishreen canals. The Nimra is about 37.8 ha (90 feddan). The distance between Abu Ishreen canals is 292 m. The water is diverted to lateral courses called Abu Sita that delivered the water to the fields. Figure 3.1 and Table 3.3 illustrate the irrigation system in Gezira Scheme. The total number of Abu Sitas has been increased since more canals have been excavated (In the past the distance between Abu Sitas was 150 m but now it is 30 m).

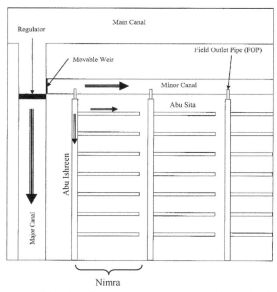

Figure 3.1. Layout of the distribution system in Gezira Scheme (Mahmoud, 1999)

Table 3.3. Gezira Scheme irrigation network

Canal	Number	Capacity (m^3/s)	Average width (m)	Length (km)
Main	2	354	50	261
Branch	11	25-120	30	651
Major	107	1.2-15	20	1,650
Minor	1,700	0.5-1.5	6	8,120
Abu Ishreen	29,000	0.116	1	40,000
Abu Sita	350,000	0.05	0.5	100,000

(Source: Ibrahim *et al.*, 2008)

3.1.4   Design of irrigation canals

Methods of the design of irrigation canals according to Mendez (1998) are:
- Tractive force method;
- Maximum permissible velocity method;
- Rational method;
- Regime theory method.

The regime method is the base of the design of stable (non-silting/non scouring) canals in Sudan. It consists of empirical equations based on observations on canals and rivers that have achieved dynamic stability. There may be seasonal deposition or erosion in the canals but the overall canal geometry in one water year remains unchanged (Munir, 2012). This can only occur when the sediment input to the canal matches the average sediment transport capacity. The regime theory was developed by Kennedy (1895) to aid the design of major irrigation systems in India followed by Lindley (1919). Lacey (1930) published the most popular set of equations. The equations were based on data from three canal systems of the Indian Sub-continent. They specify the cross-section and slope of regime canals from the incoming discharge and a representative bed material size. They are, with minor changes in the coefficients and some redefinition of the silt factor, still widely used.

The main, branch and major canals in Gezira Scheme were designed as regime conveyance canals. Matthews (1952) carried out a study in the Gezira canals to investigate the applicability of the Lacey regime equation in the main and major canals. He selected 39 canal reaches in the study. The investigations were based on the assumption that the Lacey equations would be applied and that the dimensions of stable canals would be developed from known stable canals. The study concluded that the Lacey equations are applicable in the design of the main and major canals by giving Lacey's silt factor the value $f = 0.63$. Table 3.4 summarizes the Lacey constants and the Manning coefficient (n) for the design of stable canals in Gezira. Gismalla (2009) stated that the general equations of the regime method are:

$$P = K_p Q^{1/2} \hspace{6cm} (3.1)$$

$$A = K_a Q^{5/6} \hspace{6cm} (3.2)$$

$$S_0 = K_s Q^{-1/6} \hspace{6cm} (3.3)$$

Where:
P   = wetted perimeter (m)
Q   = discharge (m^3/s)
A   = cross-sectional area (m^2)
S_0 = bed slope (cm/km)
K_p, K_a, K_s = constants (-)

Table 3.4. Different constants for the design of stable canals in Gezira

Parameters	Main canal 10 reaches	Major canal 29 reaches	All canals
N	0.021	0.017	0.018
K_p	4.55	5.51	5.26
K_a	2.75	2.6	2.64
K_s	14.57	13.9	14.5

(Source: Gismalla, 2009)

The discharge in the formula was taken as the average maximum authorized discharge. The constants dependent on the nature and magnitude of the sediment transported as well as the materials forming the canal bed and banks. The Ministry of Irrigation established the 'Design Sheet File' hydraulic design procedures for the standardized canals and hydraulic structures accordingly. The minor canals were designed for night storage water flowing continuously from the majors at night, based on the Manning equation (Plusquellec, 1999).

3.1.5 Hydraulic structures

The hydraulic control structures were designed to maintain a constant upstream level and discharge. They are controlled by manually operated means (Plusquellec, 1999). Table 3.5 describes the different types of hydraulic structures in Gezira Scheme. The main types of structures that are related to this study will be discussed.

Table 3.5. Different types of structures in the irrigation network of the Gezira Scheme

Type of structures	Range of size (m)	Total number	Function
Moveable weir	0.30 - 3.00	885	Head and cross regulator
Roller sluice gate	1.00 – 4.00	137	Head and cross regulator
Well head regulator	0.35 – 1.24	420	Head regulator
Well head regulator	0.35 – 1.24	1707	Cross structure
Night storage weir (circular/rectangular)	0.24 – 1.24	1426	Cross structure
Field outlet pipe (FOP)	0.35	28910	Head regulator

(Source: HR Wallingford, 1991)

Sluice gate regulators

The flow through sluice gates is estimated from calibration charts, which requires readings of gate opening, upstream and downstream levels (World Bank, 2000). It regulates the flow at main, branch and major canals. These types of regulators are preferable for discharge regulation as canal head regulators. Two sluice gates were installed at the offtake of Zananda Major Canal (the case study of this research) as shown in Figure 3.2 since they are less sensitive to water level change. The basic formula for submerged flow is as follows (Ahmed *et al.*, 1989):

$$Q = C_d A \sqrt{H} \qquad\qquad\qquad (3.4)$$

Where:
Q   = discharge (m³/s)
C_d = discharge coefficient (-)
A   = gates opening (m²)
H   = head difference between upstream and downstream (m)

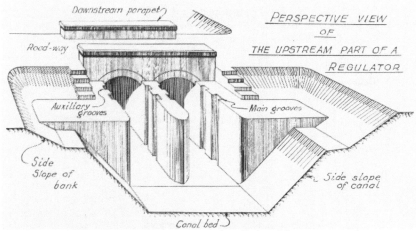

Figure 3.2. Two sluice gates at the offtake of the major canal
(Source: Design Sheet File)

Movable weir

Movable weirs are useful for regulation since the impacts of change in the crest position are more intuitively understood than for gates (Clemmens, 2006). The first description of this type of regulator was published in 1922 by Butcher, after whom the structure has been named. The weir consists of a round-crested movable gate with guiding grooves and a self-sustaining hand gear for raising and lowering it. The cylindrical crest is horizontal, perpendicular to the flow direction as illustrated in Figure 3.3. A staff gauge is attached to the weir at 0.75 of the maximum upstream water depth.

The movable weirs were installed as head and cross regulators in the major canals and most head regulators in most of the minor canals for discharge up to 5 m³/s. Table 3.6 presents the two types of movable weirs in Gezira Scheme; movable weir series-1 (MW-I) and series-II (MW-II). The weirs are designed to pass the maximum full supply level (FSL) discharge and maximum head over the crest level. For practical purposes Equation (3.5) holds good for submergence up to 70%. The discharge is reduced according to the degree of submerge as shown in Figure 3.4 therefore, a correction should be made. The degree of submergence is defined as h_2/h_1 where h_1 is the upstream water depth and h_2 is the downstream water depth over the crest of the weir. In modular flow (free flow); the flow and water levels upstream are not affected by changing the flow condition downstream whereas in drowned or submerged structures they have an effect.

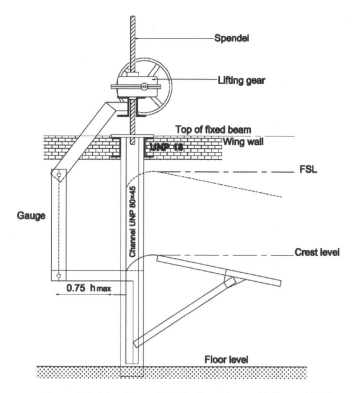

Figure 3.3. Movable weir in Gezira Scheme (Aalbers, 2012)

Table 3.6. Characteristics of the movable weir (MW)

Description	MW-I	MW-II
Range of discharge (m³/s)	< 1	1-5
Travel distance (m)	0.56	0.84
Floor level to crest level (full open) (m)	0.8	1.1
Maximum width (m)	1.3	3
Maximum depth (m)	0.6	0.8
FSL to floor level (m)	1.3	1.9
Discharge coefficient (-)	2.18	2.3
Overall depth of wall	1.5	2.2

(Source: Design Sheet File)

$$Q = C_d wh^{1.6} \tag{3.5}$$

Where:

Q	= discharge (m³/s)
w	= width of crest (m)
h	= water depth over crest level (m)
C_d	= discharge coefficient (-)

A canal with weir cross-section regulators has less in-line storage capacity than with gated cross regulators, has shorter travel and response times, and better use at long canals to improve the response of the system.

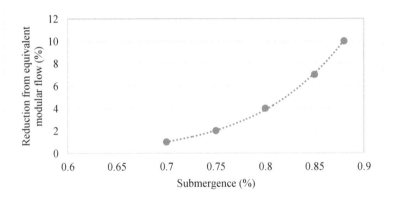

Figure 3.4. Reduction in discharge for different degrees of submergence (Ahmed et al., 1989)

Night Storage Weir (NSW)

The Gezira Scheme was originally designed to be operated as continuous irrigation system, but due to practical difficulties in irrigation at night, the system of irrigation at the minor canals was changed to be operated on what is called the night storage system in the early 1930s (Ahmed *et al.*, 1989). The types of night storage weirs (NSW) used as cross structures are rectangular and circular weirs and designed to be used as cross structures. They contain gates treated as an orifice; these gates are closed when the night storage system is in use and only opened at emergency. The idea behind the night storage system is to store water during the night by closing the gates of the field outlet pipes of the Abu Ishreen canals and the gates of the cross structures along the minor canal at 6:00 pm and release them at 6:00 am. At night the water level increases gradually along the reaches to about 20 cm above the full supply level (maximum water level) and flows from upstream to the next downstream reach over the crest of the weir to give better command for irrigation during the day. After intensification, this system is diminished gradually and changes into a continuous system. Farmers open field outlet gates 24 hours to keep pace with intensification and to cope with the deterioration of the water supply attributed to the poor maintenance of the canals (Plusquellec, 1999).

The stage-discharge equation for a broad-crested weir with a rectangular cross-section and free flow conditions ($h_2 < 0.7h_1$) can be written as Equation (3.6) (Bos, 1989), since the width of the brick wall is 0.3 m and the water over the crest not exceeds 0.2 m. Figure 3.5 displays the flow over the weir.

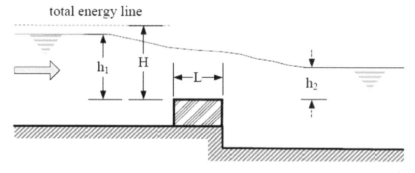

Figure 3.5 Flow over a weir (Paudel, 2010)

The discharge that passes over the wall is governed by the following equation:

$$Q = C_d w \frac{2}{3} \sqrt{\frac{2}{3} gH^3} \qquad (3.6)$$

Where
Q   = discharge (m³/s)
w   = crest width (m)
H   = total upstream head (m)
g   = gravity acceleration (m/s²)
C_d = discharge coefficient (-)

Since it is difficult to measure the total head (H), Equation (3.6) can be written as:

$$Q = C_d C_v w \frac{2}{3} \sqrt{\frac{2}{3} gh_1^3} \qquad (3.7)$$

Where h_1 is upstream water depth above crest level (m).

A round nosed rectangular broad crested weir is given by (Bos, 1989):

$$C_d = 0.93 + 0.1 \frac{H}{L} \qquad (3.8)$$

Where L is crest length at the flow direction (m).
The velocity coefficient C_v is given by (Bos, 1989):

$$C_v = (\frac{H}{h_1})^{1.5} \qquad (3.9)$$

For earthen irrigation canals $C_v < 1.05$ in most cases. The general form of Equation (3.7) is:

$$Q = Cwh_1^{1.5} \quad \text{for} \quad \text{L/h} \geq 1.5 - 3 \qquad (3.10)$$

Where C is a coefficient found in the range between 1.59 and 1.79. A value of 1.7 has been used widely with broad crested weirs (Paudel, 2010).

3.1.6 Cropping pattern

The Gezira Scheme was established with the primary objective of producing cotton. Thereafter sorghum, groundnut and wheat were introduced into the scheme. Vegetables are grown all over the year. Over the last 70 years, several changes in the cropping pattern and course rotations have been implemented as shown in Table 3.7. Since 2005 the farmers have the right to choose the cropping pattern according to the 2005 act. Figure 3.6 summarizes the cropping calendar for the main crops in Gezira Scheme for the 2007/2008 season.

Table 3.7. Cropping patterns over the years in Gezira Scheme

Season	Details of the rotation	Cropping intensity
1925 – 1930	Cotton – Dura/Lubia – Fallow – Cotton – Dura/Lubia – Fallow	(6 – Course) 66.6%
1931 – 1932	Cotton – Fallow – Fallow – Cotton – Fallow – Fallow	(6 – Course) 33.3%
1933 – 1960	Cotton – Fallow – Dura – Lubia/ Fallow – Fallow – Cotton – Fallow – fallow	(8 – Course) 50%
1961 – 1974	Cotton – wheat – Fallow – Cotton – Lubia – Groundnuts – Dura/Philipasara – Fallow	(8 – Course) 75%
1975 – 1991	Cotton – Groundnuts –Dura/Vegetable – Fallow	(4 – Course) 80% or 75% in Gezira 100% in Managil
1992 – 2005	Cotton – wheat – Groundnuts/Dura/Vegetable – Fodder – Fallow	(5 – Course) 75%

(Source: Sudan Gezira Board records)

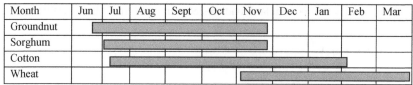

Month	Jun	Jul	Aug	Sept	Oct	Nov	Dec	Jan	Feb	Mar
Groundnut										
Sorghum										
Cotton										
Wheat										

Figure 3.6. Cropping calendar of the main crops for 2007/2008 (Mamad, 2010)

3.2   Operation and water management (past and present)

The Gezira Scheme is managed by the Sudan Gezira Board (SGB) as a government enterprise. The irrigation in the scheme is based on the demand system. Users can obtain the water directly from a supply point (FOP) with one week fixed duration. The upstream control systems are supply oriented, have limited flexibility and require agency management.

SGB used to work closely with the Ministry of Irrigation and Water Resources (now its name is Ministry of Water Resources and Electricity) that is responsible for the operation, maintenance and management of the Sennar Dam and the main, major and minor canals since the construction of the scheme in 1925. The SGB also collaborates closely with the Agricultural Research Corporation (ARC), the Sudan Cotton Company (SCC), and the Gezira State Government. In 2005, the SGB established a new act for managing the scheme 'The Gezira Scheme Act of 2005' and attempted to activate this act in 2010. Then the SGB was assigned all responsibility for the operation and maintenance of the canals. According to the 2005 act, SGB has established 1700 water user groups according to the number of the minor canals, which were named Water Users Association (WUA). The farmers are participating in the operation and maintenance of the minors and the field canals. One of its responsibilities is to collect the fees according to the cropped area. The share of the SGB is 20% of the collected fees, while the other expenditures of the maintenance activities are met by the WUA.

According to the Regulation Handbook (1934) the irrigated area in Gezira is organized into two divisions, each with three sub-divisions. On the agricultural side, it is divided into five groups. Each group has between 4 to 10 blocks (the total number of blocks is 35). To improve the level of service of this system SGB accommodated a water operation plan that is more responsive to the demand by adopting the indent system. The required discharge (indent) depends upon the duty and is rendered weekly by the sub-division engineer on Tuesdays with minor adjustment on Saturdays. The sub-division engineer passes the required amount to the next upstream sub-division engineer in the system with corrections for canal conveyance losses until the total is passed to the headwork of Sennar Dam where the headwork gates are adjusted to give the required discharge. Consequently, the water is supplied by fixed discharge for one week that means that the water level at the main canal is maintained within a certain range of fluctuation. This system is now not being adopted in practices. This method of indenting has broken-down due to difficulties in communication between the block inspectors and mismanagement of the scheme. The water duty (is number of hectares of land irrigated by cubic metre per day) gives only an approximation for allocating water and may result in a plan far from the crop needs.

3.3   Sedimentation in Gezira Scheme

The records of the Ministry of Irrigation and Water Resources (MOIWR) depict that between 1933 and 1938 the mean sediment concentration entering the Gezira Main

Canal in August was only 700 ppm, while the average sediment concentration in August of 1988 and 1989 increased to 3,800 ppm; an increase of more than five times (HR Wallingford, 1990a). According to MOIWR records the increase in sediment concentration continued to about 7,900 ppm in July 2003. The increase of sediment concentration indicates serious land degradation and soil erosion in the river basin of the Blue Nile. Poor land use practices, improper management systems and lack of appropriate soil conservation measures have played a major role of land degradation in the upstream of the Blue Nile River Basin (Setegn et al., 2009).

The continuous increase in sediment amount in irrigation canals and water mismanagement, in addition to poor government and farmers' investment in the agricultural sector are the main factors responsible for reduction of the productivity in Gezira scheme (Elhassan and Ahmed, 2008). Gesmalla (2009) stated that the average annual sediment that is entering the scheme is about 8.5 million tons. There is a great variation in the estimates of the annual sediment deposition in Gezira Scheme since it is based on estimation and not on authentic studies. Elhassan and Ahmed (2008) pointed out that the annual amount of sediment deposition in the irrigation canals is about 16 million m^3 while El Monshid et al. (1997) reported that annually 19 million tons of sediment accumulated in Gezira Scheme, but the accuracy of the values is questionable while the authors did not mention the measuring methods that they used. The sediments that accumulate in the head reach and along the canals create water delivery difficulties.

Operation of the hydraulic structures is influenced by the sediment deposition in the canals. This concerns especially movable weirs, which are sensitive to the fluctuation of the water levels. Due to this it is becoming difficult to maintain the intended discharge into the minor canals. Sometimes there is direct illegal withdrawal from the minor canals to the fields, which increases the stress of water supply downstream.

3.4   Sediment properties in Gezira Scheme

Joint efforts were carried out by HR Wallingford and MOIWR in 1988. They indicated that about 97% of the sediment entering the scheme during the flood season consists of silt and clay. This means that the sediment in Gezira Scheme is cohesive sediment. The study also concluded that the approximate distribution of sediment deposition is: 5% in the main canals; 23% in the branch and major canals; 33% in the minor canals, and 39% passed to the farm fields.

3.5   Maintenance activities in Gezira Scheme

In earlier years, when the irrigation canals were in better condition, removal of 5 to 7 Mm^3 of sediment annually was considered to be satisfactory (World Bank, 2000). In the last ten years the canal condition has deteriorated to an extent that they failed to satisfy the crop water requirements (Gesmalla, 2009). In 1999, a substantial canal desilting program was carried out. According to the records 41.0 Mm^3 of sediment was removed from the Gezira Scheme canal systems (there is over estimation of the sediment removal in addition to over digging of the canals), with a total cost of US$ 26 million (Ahmed et al., 2006). Figure 3.7 illustrates the amount of silt removal and the cost of the removal between 1987 and 2009. Excessive sediment removal in 1999 caused over digging to most of the Gezira canal system.

There is a variation of the sediment load entering the Gezira Main Canal over the years, but the general trend is increasing. The cost of sediment removal has become a major item in the MOIWR annual budget. Most of the canal cross-sections are over digged and this improper excavation even leads to large undulation of the canal beds and change in physical and hydraulic properties of the canals, which accelerates the rate of sediment deposition (Hussein *et al.*, 1986). Besides that, the clearance work is not according to the actual requirements, but depends on the availability of budget, machines, and priorities are given to the worst conditions.

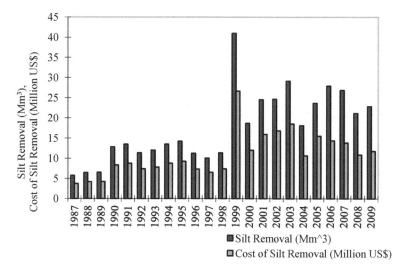

Figure 3.7. Amount of silt removal and the cost of the removal (1987-2006)

3.6   Previous studies on Gezira Scheme

HR Wallingford (1990a) conducted an extensive field study on the transport and deposition of silt in the Gezira Scheme. The study revealed that 5.9 and 4.5 million tons of sediment entered the irrigation system in 1988 and 1989 respectively. About 60% of the sediment entering Gezira Scheme settled in the irrigation canals. The highest rates of siltation occurred in the minor canals with 70% of the sediment deposition occurring over a short period from the end of July till the end of August. Most of the sediment settled at the head of the first reach of the major and minor canals..

The HR Wallingford (1990b) sediment management study was carried out for controlling sedimentation in Gezira Scheme. The study proposed sediment exclusion through gate closures of the offtake gates of the main canals at the period of high sediment load (would only be feasible with major disruption to irrigation supply) and trapping sediment in settling basins. The settling basin options do not give a complete solution to the siltation problem but only reduced the desilting campaigns to about 16% in the minor canals, according to the size suggested in the study. Furthermore, the proposed option of a settling basin at K57 was not easy to implement since its length would become about 16 km. They also proposed to construct settling basins at the head of the first reach of the major canals. This would eliminate the maintenance activities at certain locations, but the removal of sediment would be costly and the dredged sediment might create environmental problems in the scheme.

Mahmoud (1999) carried out a research in order to propose options for modernization of the Gezira Scheme. The study concluded that no sediment deposition would occur in the canal system when the system would be operated 24 hours and all the sediment would be delivered to the fields. The study referred to the deposition of the sediment in the main and major canals as a result of the night storage system in the minor canals and proposed several sediment control measures. In this study, the cohesive sediment in Gezira Scheme was not taken into consideration.

Elhassan and Ahmed (2008) estimated the reduction of the irrigated agricultural area, the decline in crop production and yield of the two main cash crops (cotton and groundnuts) in Gezira Scheme during 1970 - 2002 as a result of irrigation problems attributed to the increase in sedimentation rate in Roseires Reservoir and in the irrigation canals. The results of the regression analysis show that the cotton area decreased by 1.72% with 10% increase in sediment amount, while the area cultivated with groundnuts decreased by 0.2% with 10% increase in Roseires Reservoir sedimentation.

Mamad (2010) assessed the land and water productivity of Gezira Scheme for the growing season 2007/2008 by using remote sensing techniques. The study pointed out that land and water productivity of Gezira Scheme are low and presented clear spatial variability within the scheme. Thiruvarudchelvan (2010) assessed the irrigation efficiency in Gezira Scheme by using the same techniques and shows low irrigation efficiency. Both studies show that satellite images can be used to describe the system performance of the irrigation scheme but the accuracy and efficiency of the result is affected by the quality of the input data. Therefore, to improve the accuracy; images with high resolution and intensive ground truth were recommended.

3.7   Concluding remarks

Gezira scheme is suffering from water and sediment management problems, which increase the clearance cost. The type of the sediment is cohesive sediment since most of the sediments are silt and clay. More, elaboration will be given to the cohesive sediment transport in the next chapter. Although numerous studies have been conducted in Gezira Scheme dealing with sedimentation problems, most of them either quantify the sediment in canals or assess the performance of the scheme. There is still a need for research dealing with operation and maintenance for better sediment and water management to reduce the effect of deposited sediment in the irrigation canals.

4 Cohesive sediment transport

Cohesive sediment is defined as sediment with grain size less than 0.063 mm. It is different from sand in being cohesive. The sediment is not presented as individual particles but rather as aggregates or flocs. The formation of flocs depends on the sediment concentration, the level of turbulence, organic matter and physico-chemical properties of the sediment. The effective density of flocs is low as the flocs contain a mixture of sediment and water. In this chapter, the transportation and deposition of sediment are addressed. The sediment transport modes and factors that affect the sediment transport will be discussed. The major differences between cohesive and non-cohesive sediment will be highlighted here to give a clear picture about the type of sediment in this study. Moreover, the cohesive sediment transport processes and sediment transport under non-equilibrium conditions will also be addressed. An overview of cohesive sediment transport models will be highlighted. The numerous previous studies are elaborated.

4.1 Sediment transport modes

Sediment transport occurs in two main modes: bed load and wash load. The bed load is travelling immediately above the bed and is supported by inter-granular collision rather than fluid turbulence. It includes mainly sediment transport of coarse or fine materials (Camenen and Larson, 2005). The suspended load is the part of the bed material load which is primarily supported by the fluid turbulence and contains the portion of wash load. There is no sharp distinction between wash load and suspended load. Many engineers assumed that the sediment load consisting of sizes less than 0.063 mm (the divide point between sand and silt) is considered as wash load. The wash load is found in suspension and its transport is a function of the source of the sediment (the rate at which these particles are supplied from the river basin and not on the local transport capacity). It is easily carried in large amounts by the flow. The sum of the bed material load and the wash load becomes the total sediment load; Figure 4.1 illustrates the components of the total sediment load. Most of the sediment transport equations predict the bed material load such as the equations of Brownlie (1981), Engelund and Hansen (1967) and Ackers and White (1973). The total sediment load can be predicted only if the wash load is estimated empirically, or by analytical relations (Simon and Sentruk, 1992). Numerous relationships of transport rate and flow quantities have been proposed, many empirical and some theoretical. The empirical relations are generally applicable to limited circumstances.

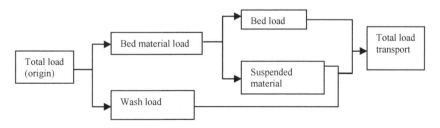

Figure 4.1. Components of total sediment load (Lawrence et al., 2001)

4.2   Factors affecting sediment transportation and deposition

According to Simons and Senturk (1992) the amount of material transported, or deposited in the stream under a given set of conditions is the result of two groups:

- sediment brought down to stream: size, settling velocity, specific gravity, shape, dispersion and cohesion. Others related to the quantity; geology and geography of the river basin; intensity, duration, seasonal rainfall, soil condition, vegetation cover and erosion. All the factors related to quantity will be addressed within the Blue Nile Hydro solidarity project through other researches;
- capacity of channels to transport sediment: canal geometry and hydraulic properties: slope, roughness, hydraulic radius, discharge, velocity distribution and fluid prosperities.

Sediment transport in irrigation canals is affected by additional factors such as the control structures, side bank influence on the velocity distribution, operation rules, water delivery schedules and many other factors.

4.3   Sediment classes

Sediments have been distinguished into two broad classes: coarse or cohesionless and fine or cohesive. The division has been arbitrarily placed on the grain size distribution. There is no clear boundary between cohesive and non-cohesive. In general, finer sized grains are more cohesive. Huang et al.(2008) clarified that sediment sizes smaller than 2 μm (classified as clay) are generally considered as cohesive sediment and greater than 60 μm as non-cohesive sediment. The sediment in between is classified as silt, considered to be between cohesive and non-cohesive sediment according to the percentage of clay. Partheniades (2009) defined cohesive sediment grains, which range in size from 1 to 50 μm.

In this study the focus will be on cohesive sediment since the sediment feature of Gezira Scheme is described as cohesive sediment according to HR Wallingford (1990a), Lawrence and Edmund (1998) and Gesmalla (2009). There is a great variation of percentage of clay that enters Gezira Scheme according to literature, but all granted that the larger portion is silt. Yosif (1993) conducted a study to determine the settling velocity of sediment in turbulent flow in Gezira Scheme, the study found that D_{95} is 25, 24 and 21 μm in the main, major and minor canals respectively.

The coarse sediment transport in appreciable quantities in the bed and their rate of transport is a function of flow conditions (Simons and Senturk, 1992). Most of the coarse sediments are transported as bed load. Coarse grains in suspension behave independently from each other with exception in high dense suspensions. In contrast, fine sediments are encountered in the bed in only very small quantities compared to the total load and transport in suspension with their transport rate related to their supply rates. Partheniades (2009) stated that fine sediments are subjected to a set of attractive and repulsive forces of electrochemical and atomic nature acting on their surfaces and within their mass. These forces are the result of the mineralogical properties of the sediment and of the adsorption of ions on the particle surfaces.

The fine sediment grains have in general a flat plate or a needle shape and a high specific area (Partheniades, 2009). High concentrations of fine sediment might transform the water flow into a mud flow. Many pollutants, such as heavy metals, pesticides, and nutrients preferentially adsorb to cohesive sediments.

4.4   Cohesive sediment processes

Some features that characterize cohesive sediment will be described in this section.

4.4.1   Aggregation

The rate of fine sediment aggregation is related to the frequency of particle collisions, which is a function of fluid flow and particle characteristics plus the concentration of particles (McAnally and Mehta, 2000). Metallic or organic coatings on the particles may also influence the inter-particle attraction of fine sediments. The cohesive sediment particles have strong inter-particle forces due to their surface ionic changes, which causes the small particles to bind together to form larger flocs (Huang *et al.*, 2008). The flocs may collide with other particles of flocs or may break up by turbulent stress. The interaction between particles will result in floc formulation or dispersion. The distribution of sizes and settling velocities of the flocs depend upon the physico-chemical prosperities of the sediment water system and flow parameters (Partheniades, 1986). Huang *et al.* (2008) stated that many formulae that determine the settling velocity of cohesive sediment such as Nicholson and O`Connor (1986), Burtan *et al.* (1990) and Van Leussen (1994). Krone (1962) found that the settling velocity increases with the sediment concentration and proposed the following formula:

$$w_s = kC^n \qquad\qquad\qquad\qquad (4.1)$$

Where:
w_s = settling velocity (m/s)
C   = suspended sediment concentration (g/l)
k   = empirical constant (-)
n   = an exponential (-)

Krone (1962) set n = 4/3, k = 0.001, while Cole and Miles (1983) set k between 0.001 - 0.002 and n = 1. Many other formulas are presented in Huang *et al.* (2008).

Soil aggregates are cemented clusters of sand, silt, and clay particles. Clay particles have a negative electrical charge and repel each other. A cation has a positive charged molecule and can render clay particles stick together (flocculate).

4.4.2   Deposition

Deposition of cohesive sediment is a very complex process, since particle shape and size (and eventually fall velocity) change due to particle flocculation and aggregation (Scarlatos and Lin, 1997). Under certain conditions, the attractive forces exceed the repulsive ones so that colliding particles stick together and form agglomerates known as flocs. The flocs are varying with size and settling velocities. Their settling velocities are much higher than those of the individual particles (Partheniades, 2009). Rapid deposition can then take place. This phenomenon is known as flocculation. The deposition of cohesive sediment is controlled by the bed shear stress, turbulence processes in the zone near the bed, settling velocity, type of sediment, depth of flow, suspension concentration and ionic constitution of the suspending fluid (Partheniades, 2009). Deposition occurs when the bottom shear stress is less than the critical shear stress. Krone (1962) developed the following formula:

$$D = w_s \, C_b (1 - \frac{\tau}{\tau_d})$$ (4.2)

Where:
D   = deposition flux (kg/m^2/s)
C_b = near bed sediment concentration (kg/m^3)
w_s = settling velocity (m/s)
τ = bed shear stress (N/m^2)
τ_d = critical shear stress for full deposition (N/m^2)

Critical shear stress of deposition

Krone (1962) started the investigation of cohesive sediment transport by using San Fransisco bay sediment and identified values of critical shear stress for the onset of movement, beyond which the particles cannot settle. He gave three different deposition laws. For the concentration up to 300 ppm, the various order aggregates settle almost independently without much mutual interference. For concentrations between 300 and 10,000 ppm, the collision increases and results in larger aggregates due to higher collision frequency with higher settling rates and develops stronger bonds with the bed material. The values of critical shear stress for deposition are given in Table 4.1. For a higher concentrations than 10,000 ppm, fluid mud is assumed to settle more slowly than in previous cases, which is known as hindered settling. Partheniades (1965) concluded from his experiments that sedimentation and erosion cannot occur simultaneously in cohesive sediment dynamics. Mehta and Partheniades (1973) carried out experiments with San Francisco Bay mud, Lake of Maracaibo mud and processed kaolinite clay. They stated that not all the sediment deposited when the bed shear stress is less than the critical stress for deposition and an equilibrium concentration was found. The equilibrium sediment concentration is the concentration of relatively weak flocs that do not have sufficiently strong bonds and will be broken down before reaching the bed, or will be eroded immediately after being deposited (Winterwerp and Kesteren, 2004). Partheniades (1977) reported that if the bed shear stress is between the critical shear stress of erosion and critical shear for deposition, no sediment fraction exchanges between water and bed and this yield the condition of wash load. The very fine material, such as organic matter and colloidal particles cannot settle because of Brownian motion and their excess density of particle is too small. McCave and Swift (1976) suggested that the critical stress of deposition (τ_d) is probably related to the diameter, or more properly to the settling velocity of the particle. They assumed that τ_d is given by critical shear stress of erosion for non-cohesive sediment since below that value movement occurs and any grain reaching the bed would stick. A relation has been developed based on measurements in the laminar flow cell:

$$\tau_d = (S-1)\rho_w d$$ (4.3)

Where:
τ_d = critical shear stress for full deposition (N/m^2)
S   = specific gravity of sediment (-)
d   = particle diameter (m)
ρ_w = density of the water (kg/m^3)

   The general values of critical shear stress of deposition used in modelling studies are between 0.05 and 0.1 N/m² (Whitehouse *et al.*, 2000, Milburn and Krishnappan, 2003). Maa *et al.* (2008) stated that deposition only occurs when the local bed shear stress is smaller than a critical value.

   In contrast Winterwerp and Kesteren (2004) assumed that erosion and deposition can occur simultaneously. They suggested that wash load occurs when there is a larger potential erosion rate than the deposition flux or vice versa. The deposited particles are re-entertained immediately by turbulent flow, the assumed critical sediment deposition does not exist and the deposition rate is given by the following formula:

$$D = w_s C_b$$
(4.4)

   This assumption is in agreement with Sanford and Halka (1993) and Chan *et al.* (2006). They analysed a series of field measurements under tidal conditions in Chesapeake Bay, compared them with simulated sediment by applying continuous deposition and found acceptable results. The existence of critical shear stress for erosion is widely accepted, but for deposition it is still being debated (Maa *et al.*, 2008).

Table 4.1. Different values of critical shear stress for deposition in various studies

Studies	Critical shear stress (N/m²)	Conditions
Krone (1962)	0.06	C < 300 ppm
	0.078	C = 300 - 10000 ppm
McCave and Swift (1976)	0.04	u* = 0.67 cm/s, u = 0.15 - 0.2 m/s for 0.01 mm silt.
Dade *et al.* (1992)	0.015-0.03	u = 0.08 - 0.23 m/s
Hunt's (1986) and Ledden (2003)	0.04-0.16	for breakup of montomorillonite and illite flocs

Note: C: sediment concentration, u: flow velocity, u*: shear velocity

Settling velocity

The settling velocity of particles is a function of the particle shape, size, density and viscosity of the flow. It can be obtained from the balance between drag force and gravity force in homogeneous fluid that is known as Stokes formula for individual particles. This formula is also valid for flocs of cohesive sediment although no data on the drag coefficient for falling flocs are available (Winterwerp and Kesteren, 2004):

$$w_s = \frac{(\rho_s - \rho_w)gD_f^2}{18\mu}$$
(4.5)

Where:

w_s	= settling velocity (m/s)
g	= gravity acceleration (m/s²)
μ	= eddy viscosity (kg/m/s)
ρ_s	= particle density (kg/m³)
ρ_w	= water density (kg/m³)
D_f	= floc diameter (m)

This theory is applicable to suspension sediment with low concentration. The flocs or particles start to hinder each other in their settling when the concentration increases. The effect of the hinder appears when the concentration is greater than 10,000 ppm (Sanchez et al., 2004, Partheniades, 2009). The settling of individual particles is affected by return flow, which affects other nearest particles in the vicinity of the falling particle. The hinder has an effect in increasing the viscosity of the fluid, which decreases the effective settling (Partheniades, 2009). The increase of the settling velocity with concentration is a clear indication of flocculation effects (Vinzon and Mehta, 2003; van Rijn, 2007).

Richardson and Zaki (1954) developed the semi-empirical Equation (4.6). The exponent n in Equation (4.6) is found in the range $2.5 < n < 5.5$ and it is a function of particle Reynolds number (Re_p). Figure 4.2 shows the variation of n with Reynolds number of the particle.

$$W_s = w_s(1-\phi_s)^n \qquad\qquad (4.6)$$

$$Re_p = \frac{w_s d_{50}}{\gamma} \qquad\qquad (4.7)$$

Where:
W_s = effective settling velocity (m/s)
w_s = settling velocity (m/s)
n   = exponent (-)
d   = particle diameter (m)
v = kinematic viscosity (m²/s).
ϕ_s = volumetric concentration (-) $\phi_s = c/\rho_s$, C is sediment concentration (kg/m³)
ρ_s = particle density (kg/m³)

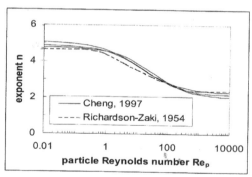

Figure 4.2. Variation of n with Re_p according to Richardson and Zaki (1954)
($\Phi_p = 0.05$, 0.1 and 0.4) and Cheng (1997)(Winterwerp and Kesteren, 2004)

Slaa et al. (2012) studied the sedimentation behaviour of silt in the hindered settling regime by conducting laboratory experiments. The result was compared with the widely used Richardson and Zaki (1954) formula. They concluded that the Richardson and Zaki formula is under estimating the settling velocities for the finest silts at high concentrations, but performed well with coarsest silt. They referred this to the effect of particle size on the effective viscosity, which was not taken into account in the Richardson and Zaki formula. When the result was compared with Equation (4.8) of Winterwerp it gives a better result for high concentrations and fine silt.

$$W_s = w_s \frac{(1-\phi)(1-\phi_s)}{1-2.5\phi}$$ (4.8)

Where ϕ is the ratio between sediment concentration and gelling concentration, which is computed with the flocculation model, or is obtained from measurements. Winterwerp and Kesteren (2004) defined the gelling concentration as the concentration for which the flocs form a space-filling network.

Vertical distribution of suspended sediment

Fluid turbulence has an essential role in the suspended sediment near the bed (McCave, 2008). To maintain an equilibrium profile, the upward sediment flow due to diffusion must be balanced by the quantity of settling sediment (McCave, 2008):

$$w_s C + \varepsilon_s \frac{dC}{dy} = 0$$ (4.9)

Where:
C = time average concentration (kg/m^3)
w_s = settling velocity (m/s)
ε_s = eddy diffusivity of sediment (m^2/s)

By assuming that the vertical profile of eddy diffusivity is parabolic and the vertical velocity is logarithmic, the linear distribution of shear and the sediment fluid interaction don't play a role (Winterwerp and Kesteren, 2004). In that case the equilibrium solution of Equation (4.9) as presented by Rouse (1937) is known as the suspended sediment distribution equation:

$$\frac{C}{C_a} = \left[\frac{d-y}{y} \frac{a}{d-a} \right]^{z_1}$$ (4.10)

$$z_1 = \frac{w_s}{\beta k_r U_*}$$ (4.11)

Where:
C_a = reference concentration at level a (kg/m^3)
y = total depth (m)
a = reference depth from bed (m)
d = water depth from bed level (m)
z_1 = Rouse number (-)
k_r = Karman number equal to 0.41 (-)
U_* = shear velocity (m/s)
β = turbulent- Schmidt number (-)

The equation can be written in the following form:

$$\frac{C}{C_a} = \left[\frac{\eta - \eta_a}{1 - \eta_a}\right]^{z_1} \text{ Where } \eta_a = \frac{a}{d} \text{ and } \eta = \frac{a}{y} \qquad (4.12)$$

The vertical concentration distribution has been investigated by many researchers and all show a satisfactory agreement with the form of Equation (4.12). Brush et al. (1962), Matyukhin and Prokofyev (1966) and Majumdar and Carstens (1967) conducted extensive experimental works and they concluded that for fine sediment $B = 1$ (when the diffusion coefficient of the sediment equals the kinematic eddy viscosity or the diffusion coefficient for momentum). However, the exponent z_1 was obtained by graphically fitting the data and not from theory (Simons and Senturk, 1992).

4.4.3 Consolidation

Two types of consolidation are usually considered: primary and secondary. Primary consolidation is caused when the self-weight of particles expels the pore water and forces the particles closer together. Primary consolidation ends when the seepage forces have completely dissipated. Secondary consolidation starts during the primary consolidation. It is caused by the plastic deformation of the bed under a constant overburden and may last for weeks or months (Huang et al., 2008). The consolidated sediment requires a high rate of shear before re-suspension.

4.4.4 Erosion

Erosion and deposition of bottom sediments reflect a continuous, dynamic adjustment between the fluid forces applied to a sediment bed and the condition of the bed itself (Sanford, 2008). The factors that affect the erosion indirectly affect the deposition. The physico-chemical properties, biological properties and physical properties are dominantly affecting the erodability. The physical properties that affect the erodability are clay content, water content, clay type, temperature and bulk density (Huang et al., 2008). The critical shear stress of erosion was shown to be representative of the physico-chemical properties of the sediment and water system. Millar and Quick (1998) defined the erosion as individual particles or small aggregates that are removed from the soil mass by hydrodynamic forces such as drag and lift. The erosion or re-suspension occurs when the bed shear stress exceeds the critical shear stress for erosion. The first experimental work in erosion was carried by Partheniades (1962) in a systematic way with mud from San Francisco Bay. He developed the following formula:

$$E = M \left(\frac{\tau}{\tau_c} - 1\right) \qquad (4.13)$$

Where:

E   = erosion flux ($kg/m^2/s$)
M   = erosion rate parameter or re-adaptability coefficient ($kg/m^2/s$)
τ = bed shear stress (N/m^2)
τ_c = critical shear stress for erosion (N/m^2)

The equation was developed for well consolidated and homogeneous beds. Parchure and Mehta (1985) reported that the critical stress for erosion was in the range of 0.04 to 0.62 N/m² based on laboratory analyses. Values within this range are often applied in modelling studies. Winterwerp and Kesteren (2004) gave a range for the critical stress of 0.1 to 5 N/m². The high range may relate to pure mud. Lumborg (2005) conducted tests for different types of sediment and gave 0.16 N/m² for mixed mud and 0.3 N/m² for pure mud. It increases with depth in order to represent more sediment consolidation. He reported in his study that others gave a range of 0.05 to 0.1 N/m². However, for small irrigation canals Lawrence and Edmund (1998) clarified that the bed shear is limited by small flows and depths. Therefore there is very little reworking for sediment deposits.

4.5   Sediment transport under non-equilibrium conditions

Non-equilibrium sediment transport refers to cases where the outflowing sediment from a reach does not equal the inflowing sediment to that reach while all processes of sedimentation: erosion, entrainment, transport, deposition, and consolidation are active. As the prediction of sediment transport in non-equilibrium conditions is very complicated, mathematical models are usually used. Sediment transport formulas are used in the analysis of non-equilibrium conditions although there are significant differences. Sediment discharge formulae are functional only for the bed material load and based on equilibrium conditions. In contrast, non-equilibrium models are functional for both bed material load and wash loads and can calculate both equilibrium conditions and changes in bed profile due to sediment inflow deficit or excess.

Suspended sediment motion in the fluid caused by turbulence effects is assumed to be analogous to diffusion-dispersion processes (Simons and Senturk, 1992). Equations describing the dispersion of mass in open channel flow can be easily derived considering the conservation of mass, which states that the rate of change of weight of dispersants in the control volume is equal to the sum of the rates of change of weight due to convection and diffusion. The general equation describing the dispersion process in turbulent flow can be written as (Partheniades, 2009):

$$\frac{\partial c}{\partial t} = -u \frac{\partial c}{\partial x} - v \frac{\partial c}{\partial y} - w \frac{\partial c}{\partial z} + \frac{\partial}{\partial x}\left[\varepsilon_x \frac{\partial c}{\partial x} \right] + \frac{\partial}{\partial y}\left[\varepsilon_y \frac{\partial c}{\partial y} \right] + \frac{\partial}{\partial z}\left[\varepsilon_z \frac{\partial c}{\partial z} \right] \qquad (4.14)$$

Where:

c   = time averaged concentration (ppm)
u, v, w   = time averaged velocities in x, y and z direction (m/s)
x, y, z   = directions in a rectangular coordinate system (m)
$\varepsilon_x, \varepsilon_y, \varepsilon_z$ = diffusion coefficients (m²/s)

The time averaged velocities or convection velocities are tending to move the suspended load in various directions. Models for simulation of non-equilibrium, suspended sediment transport in open canals are sub-divided into convection-diffusion, energy and stochastic models (Timilsina, 2005). The diffusion models (I D, 2D and 3D) have formed the greatest acceptance for practical application and are usually part of the numerical or mathematical models of suspended sediment transport (Raudkivi, 1990). Many developed mathematical models are solving the convection-diffusion equation. More details were given by Mendez (1998). Scarlatos and Lin (1997) developed a one dimensional model for quantification of fine-grained sediment movement in small canals. The model simulates the effects of deposition and erosion through appropriate sink and source terms, but the model assumes that the changes in the thickness of the bottom layer due to erosion and deposition do not affect the bathometry of the channel. Van Rijn (1987) developed a two dimensional vertical and a three dimensional mathematical model based on the convection diffusion equation to study the morphological process in case of suspended sediment transport. The convection diffusion equation was solved by the finite elements method. The flow velocities and sediment mixing coefficients are computed by a model based on flexible profiles in vertical direction. The vertical distribution of the fluid mixing coefficient is described by a parabolic constant profile. The two dimensional vertical model has been verified extensively by using flume and field data.

In contrast, Galappatti (1983) developed a depth integrated model based on an asymptotic solution for the two-dimensional convection equation in the vertical plane. The concentration is expressed in terms of a depth average concentration. The model describes how the mean concentration adapts in time and space towards the local mean equilibrium concentration. The validity of the model was studied by Wang and Ribberink (1986). They concluded that the model will not work for large deviation between the concentration profile and the equilibrium profile. The model is only valid for fine sediment, the factor w_s/u_* should be smaller than 1 and the recommended range is between 0.3 and 0.4. Also the time scale and length scale of flow variation should be larger than h/u_* and $v*h/u_*$ respectively, where: u_* is shear velocity, w_s is settling velocity, h is water depth and v is mean velocity.

4.6   Overview of cohesive sediment transport models

Here are some of the widely used models that are dealing with cohesive sediment:
- MIKE 11 was developed by Danish Hydraulic Institute in 1993. It is a one dimensional hydrodynamic software package. It includes the full solution of the St. Venant equations. The model has many modules; advection-dispersion, water quality and ecology, sediment transport, rainfall-runoff, flood forecasting, real-time operation and dam break modelling. MIKE 21C is an integrated river morphology modelling tool based on a curvilinear version of the water model MIKE 21 adjusted to river applications. The model system simulates transport of all sediment sizes from fine cohesive material to gravel by using a multi-fraction approach and a number of transport formulae;
- SHARC Sediment and Hydraulic Analysis for Rehabilitation of Canals is a software package that was developed by HR Wallingford in 2001. It integrates a sediment routing model with design packages for sediment control structures, and other modules. The transport capacity of the canal can be predicted by using the DORC module (Design of Regime Canals) based on the canal design. A range of canal design methods is available in the DORC module. It is possible to estimate the fine sediment concentration as well as sandy sediment. The input concentration for each fraction is compared with the transport capacity as computed by using the Westrich and Jurasak equation (Lawrence et al., 2001). The approach used in the prediction of fine sediment concentration is not widely accepted;
- ISIS-Sed is a sediment transport module used with the ISIS flow model, developed by HALCROW and HR Wallingford for studying the morphology of rivers and alluvial canals for uniform and graded sediment (Paudel 2010). It can

predict the amount of deposition and erosion and sediment transport rates, and can simulate the cohesive sediment transport;

- SOBEK is a software package based on a 1D hydrodynamic model, developed by WL/Delft Hydraulics and the Government Institute for Inland Water Management and Wastewater Treatment in 1994, for integral simulation of processes of steady and unsteady flow. It can be used in rivers, canals and sewer networks for salt intrusion, non-cohesive sediment transport, river morphology and water quality (Zuwen et al., 2003). It was recently linked to the DelWAQ model to simulate cohesive sediment but changes in the morphology are not considered;

- Delft3D package has been developed by WL/Delft Hydraulic in cooperation with Delft University of Technology. It is a model system that consists of a number of integrated modules which together allow the simulation of hydrodynamic flow, sediment transport (cohesive and non-cohesive) and morphological change (Wubneh, 2007). The model is basically developed to be applied in estuaries, coastal areas and rivers, but it has not been prepared to be adapted to systems with different types of hydraulic structures;

- SED2D WES is a 2D finite element model for cohesive sediment transport developed by the US Army Waterways Experiment Station. The model can simulate the cohesive and non-cohesive sediment, but the model considers a single, effective grain size during each simulation (Huang et al., 2008).

Several studies of cohesive sediment transport have been conducted during the last decades by using mathematical models. Most of the sediment models have been developed to solve specific types of problems. A finite element model (TABS-2) was presented by Thomas and McAnally (1985) to calculate the bed change in a channel with cohesive and non-cohesive sediment. The sediment transport equations in the model were solved based on the bed material load, which may be improper for non-equilibrium sediment transport. Celik and Rodi (1988) developed a model, but it cannot handle complex boundaries as in irrigation systems. Van Rijn et al. (1990) developed a model that considers suspended load transport. The model cannot simulate bed load transport and can only be used for non-cohesive sediment transport. Guan et al. (1998), Wu et al. (1999), Liu et al. (2002) and Lopes et al. (2006) modelled cohesive sediment transport and the dynamics of the sediment in estuaries. Liu et al. (2002) simulated hydrodynamics and cohesive sediment transport in Tanshui River estuarine system in Taiwan by using a lateral average two dimensional numerical model. The result indicted that more research is needed in this field. Related to this concern, the SOBEK model was improved by Zuwen et al. (2003) to deal with cohesive sediment. A reach of about 38 km downstream from the LJ station in the Yellow River Estuary was selected for the simulation. The model was validated by using a prototype data set to address the dredging and efficiency of dredging. The cohesive sediment processes in estuaries are different from rivers since they are affected by other factors such as salinity of the sea and tidal flows. A two dimensional depth average model for non-equilibrium non cohesive and cohesive sediment transport in alluvial canals was developed by Hung et al. (2009). The processes of hydraulic sorting, armouring and bed consolidation are included in the model. The model's applicability has been demonstrated through the application to the Ah Gong Diann Reservoir in Taiwan.

4.7   Concluding remarks

Most cohesive sediment studies have been conducted in rivers, estuaries, and along shores. So far none has simulated the cohesive sediment transport in irrigation canals. The cohesive sediment processes in estuaries are different from rivers since they are affected by other factors such as salinity of the sea and tidal flows. There was still a need to develop a model to simulate the cohesive sediment transport in irrigation canals provided with different types of control structures. A model to simulate the suspended sediment transport cannot be developed without deep understanding of the properties of the sediment under study. Thus, field measurements and data collection were conducted to calibrate and validate the developed model. More elaboration of the field measurements and scheme analysis can be found in the following chapter.

5 Data collection and scheme analysis

5.1 Conceptual approach

In order to handle the sediment problems in Gezira Scheme field measurements were performed in 2011 and 2012 from June to October, since most of the sediment is entering the scheme during this period. Irrigation canals and measurement locations were selected for the study. The discrepancy of the water level along the canals and the effect of the morphological change on the water level have been highlighted. The effect of the change of the gate setting on the stability of the water level has been analysed. The quantity of sediment deposited in the canals has been investigated though the mass balance and cross-section surveys. Furthermore, the link between irrigation system operation and resulting system performance has been developed. The relation between the actually released amount of water and the crop water requirement has been generated. The deposition process of the sediment was addressed by performing laboratory analyses. The maintenance activities during the study period have been elaborated.

5.2 Study area and measurement locations

Gezira and Managil main canals are not suffering from serious sediment deposition as in the major and minor canals. Only about 5% of the total deposition takes place in the two main canals (HR Wallingford, 1990a). A study showed that siltation in the Managil Main Canal is due to a wrong design section, since it was over designed and would have to be remodelled in order to reduce sedimentation (Gismalla and Fadul, 2006). In this study the focus is on the siltation in the major and minor canals of the scheme.

The location of the study area within Gezira Scheme is shown in Figure 5.1. Zananda Major Canal was selected to be studied in detail. It is the first canal that takes water from the Gezira Main Canal under gravity irrigation. The location of the offtake is at 14°01′42″N, 33°32′33″E. It has a total length of 17 km. The canal delivers water to 9 minor canals with a total command area of about 8520 ha. Two cross regulators, type movable weir (MW-II), were installed at 9.1 and 12.5 km from the offtake with a crest width of 2 and 1.3 m respectively. Figure 5.2 shows the scheme of the major canal under study and the measurement locations. The 6 km long Toman Minor Canal that takes water at 12.5 km from the Zananda offtake was also selected for the study. It is located at the mid of the system, accessible and in good condition. It serves a command area of about 772 ha. The area is divided into 21 Nimra. Each Nimra is irrigated by an Abu Ishreen canal as shown in Figure 5.3. A village named El-Egaida is located at the end of the Nimras 8 and 9. The canal has four reaches divided by two night storage weirs (NS) and a pipe regulator (diameter 0.76 m) at the head of the fourth reach. Tables 5.1 and 5.2 give information about the characteristics of the canals.

About 14 monitoring points have been selected at different locations. One at the offtake of Gezira Main Canal, three control points along Zananda Major Canal, at the offtake and downstream of the intermediate structures. Moreover, 7 positions at the offtake of the lateral canals at the first and second reaches of Zananda Major Canal. Three control points at cross structures at Toman Minor Canal. The locations of different types of control structures are shown in Table 5.3. Most of the minor canal offtakes are movable weirs type one (MW-I). However, weirs type two (MW-II) were installed as cross structures along most of the major canals.

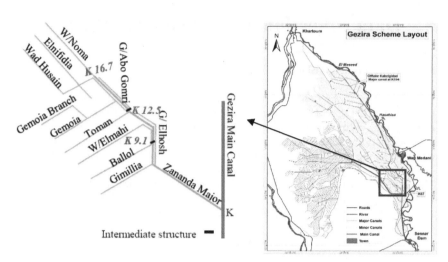

Figure 5.1. Study area of Gezira Scheme

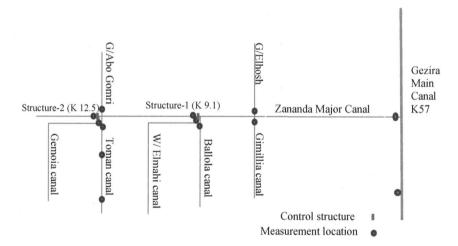

Figure 5.2. Scheme of Zananda Major Canal and measurement locations

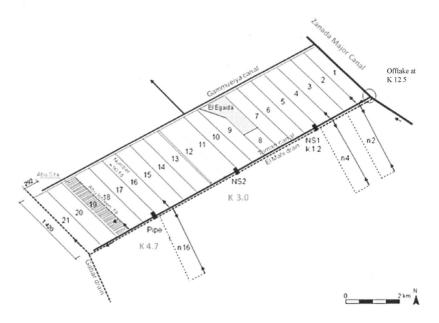

Figure 5.3. Irrigated areas of Toman Minor Canal (Aalbers, 2012)

Table 5.1. Characteristics of Zananda Major Canal

Canal characteristic	Value
Position of offtake along the main canal (km)	57
Command area (ha)	8,520
Effective length (km)	17
Number of reaches	3
Supplied minor canals	9
Design discharge (m^3/s)	3.52
Full capacity of the canal (m^3/s)	5.5

(Source: HR Wallingford, 1991)

Table 5.2. Characteristics of Toman Minor Canal

Canal characteristic	Value
Position of offtake along the major canal (km)	12.5
Command area (ha)	772
Effective length (km)	4.6
Number of reaches	4
Design discharge (m^3/s)	0.46
Head regulator: weir width (m)	0.8
Night-storage weirs	2

(Source: HR Wallingford, 1991)

Table 5.3. Different types of structures along the canals under study

Canal	Distance from offtake (km)	Location	Type of structure	Width (m)
Zananda	0	Offtake of Zananda at 57 km from Sennar Dam	Tow sluice gates	2
	7.4	Gimeliya offtake	MW-I	0.8
	7.4	G/ Hosh offtake	MW-I	0.8
	9.1	Cross structure-1	MW-II	2
	9.1	Ballola offtake	MW-I	0.8
	9.1	W/ Elmahi offtake	MW-II	1
	12.5	Cross structure-2	MW-II	1.3
	12.5	G/ Abu Gimri offtake	MW-I	0.35
	12.5	Gommuiya offtake	MW-I	1
	12.5	Toman offtake	MW-I	0.8
Toman	1.2 km from Toman offtake	Cross structure	NS-1	Diameter 0.91
	3 km from Toman offtake	Cross structure	NS-2	Diameter 0.91
	4.7 km from Toman offtake	Cross structure	PR	Diameter 0.67

Methodology of field data collection

Data of the studied canals were collected as follows:

- *water level measurement:* manual monitoring by using staff gauges and automatic monitoring instruments such as divers/loggers;
- *flow velocity:* by applying depth integrated techniques by using current-meters and Parshall flumes for the field canals;
- *sediment data:* such as sediment concentration by using the dip sampling technique and bed samples by using a bottom grab to get the grain size distribution and to study the properties (physico-chemical and mechanical properties). Grain size distribution for the suspended sediment was obtained from the samples that were collected by measuring the sediment concentration;
- *bathymetric survey:* cross-sections every 200 m;
- *other data:* cropped area by survey, sowing dates and design discharge;
- *maintenance activities and the overall condition of the field:* by report observations and interviews with farmers and operators.

Table C.1 in Annex C gives a summary of the collected data, frequency and their location. The total annual number of recorded water levels was about 1080 and about 1290 for the sediment samples that have been analysed.

5.3   Water level measurements

Water levels were measured manually by using staff gauges at the intakes and cross structures on a daily basis at the selected locations as shown in Figure 5.2. The measurements depended on the type of the structures:

- *sluice gates*: water levels upstream and downstream, as well as opening and closing of each gate were recorded;

- *movable weirs*: water levels upstream, the position of the weir crest and depth of the water over the crest has been measured. Downstream water levels have also been measured to check the submergence.

Automatic monitoring instruments such as divers/loggers were installed upstream of the inlet of Toman and Zananda canals in 2011 and 2012. Three more instruments were installed downstream of the offtake of Zananda Major Canal and Toman Minor Canal, and at the end of the first reach in 2012 to record the water levels every 2 hours.

A reasonable agreement between the manually collected data and the logger records was found for the water level upstream of the head regulator of Zananda Major Canal in 2012 as revealed in Figure 5.4.

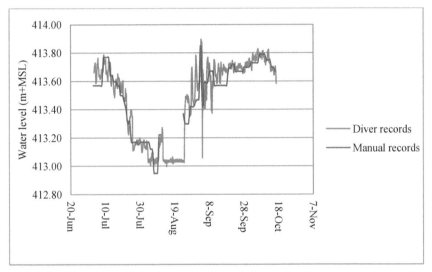

Figure 5.4. Manually collected data and logger record of the water level upstream of the offtake of Zananda Major Canal in 2012

5.3.1 Stability of the water level along the major canals

Comparison of the actual water levels with the full supply levels provides an indication of the performance of the distribution system. The analysis of the recorded water levels along Zananda Major Canal shows that they are fluctuating far above the full supply level at most of the locations. The control structures were designed to maintain a constant upstream water level at full supply level (FSL), but the actual situation is far from the design. Figures 5.5 and 5.6 show the upstream water levels at the offtake of Zananda Major Canal. Most of the head regulators of the minor offtakes were installed just upstream of the cross regulators in the major canals, such as the group regulators at 9.1 and 12.5 km from the offtake. In this way a proper head above the crest level of the offtakes can be maintained by adjusting the setting of structures to draw their fair share of water. The upstream water level at Zananda Major Canal in 2011 shows more stability than in 2012. Figure 5.7 presents the gate openings and displays the closure period (satisfy the demand due to rainfall) in 2011 and 2012. There is instability in the system, especially after the rainfall period when all the farmers irrigate their land at the same time. A great demand results into a large amount of water delivered to the system, which exceeds the full capacity of the canals. The downstream water level is influenced by changes in the upstream water level and by the gate opening, and sometime adhering to the same trend as the upstream water level as shown for the offtake of Zananda Major Canal in Figure 5.6. On 8 August 2011 the water level upstream of the offtake of Toman Minor Canal increased gradually to 1.1 m above FSL (date of the closure of the major offtake). The reason behind this was the closure of the offtakes of most of the minor canals. To achieve stability of the water level the gates of the control structures need to be adjusted when the water supply and demand are changing.

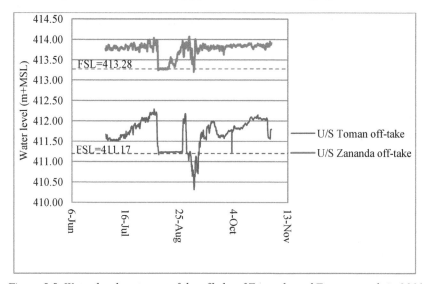

Figure 5.5. Water level upstream of the offtake of Zananda and Toman canals in 2011

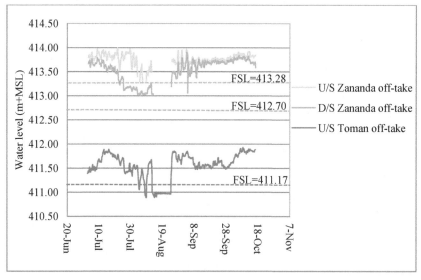

Figure 5.6. Water level at the offtake of Zananda and Toman canals in 2012

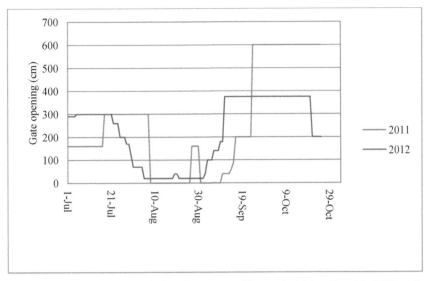

Figure 5.7 Gate setting at the head structure of Zananda Major Canal in 2011 and 2012

Table 5.4 and Table 5.5 summarize the deviation of the mean water levels above the full supply level for different locations in 2012 and 2011. The reference values for the measurements were taken at 412.35 m+MSL and 411.50 m+MSL at the top of the fixed beam of the cross structures at 9.1 and 12.5 km respectively from the offtake according to design of the canal. The standard deviation of the upstream water levels at the control structures along the major canal varied between 0.15 and 0.21 m in 2011 and 2012 with more stability upstream of K 9.1 in 2011. In 2011 downstream of the offtake a higher stability is shown than upstream (standard deviation is 0.078 m). HR Wallingford (1991) studied the variability in the water levels for the similar canal under the study in 1988 and 1989 and found that the standard deviations were 0.11, 0.065 and 0.11 m upstream of the head regulator and the cross regulators at K 9.1 and K 12.5 respectively. This indicates that the water levels were more stable at that time compared to the study period. An inspection of the water level records reveals that the variability in the upstream water levels at the offtake of the major canal is confined to the rainy period, while at other locations it extended all over the study period. The records in 2011 show high values of the mean water levels above FSL as a result of more water that was delivered to the system during the study period in that year.

Table 5.4. Variation of water level along Zananda Major Canal in 2012

Location	Max level (m)	Mean level (m)	Min level (m)	FSL (m)	Standard deviation (m)	Mean level above FSL (m)	Peak level above FSL (m)
U/S offtake at k57	413.99	413.77	413.11	413.20	0.16	0.57	0.79
D/S Zananda offtake	413.89	413.52	412.95	412.80	0.23	0.72	1.09
K 9.1	412.83	412.52	411.89	411.83	0.20	0.69	1.00
K 12.5	411.89	411.62	410.89	411.17	0.18	0.38	0.72

Note: all the levels are above MSL; D/S = downstream; U/S = upstream

Table 5.5. Variation of the water level along Zananda Major Canal in 2011

Location	Max level (m)	Mean level (m)	Min level (m)	FSL (m)	Standard deviation (m)	Mean level above FSL (m)	Peak level above FSL (m)
U/S offtake at K 57	414.00	413.75	413.61	413.20	0.17	0.55	0.80
D/S Zananda offtake	413.67	413.61	413.28	412.80	0.078	0.81	0.87
K 9.1	412.92	412.61	412.34	411.83	0.15	0.78	1.09
K 12.5	412.28	411.81	411.08	411.17	0.21	0.61	1.11

Note: U/S: Upstream, D/S: Downstream, Min: Minimum, Max: Maximum, FSL: Full Supply Level, all the levels are above MSL

5.3.2 Stability of the water level along the minor canals

The water level was measured at various locations upstream and downstream of the control structures along Toman Minor Canal (at the off take and 1200, 3000, 4700 m from the offtake). Figure 5.8 displays the water level at the head of Toman Minor Canal between 7 September and 16 October 2012 for the first reach. The change in water level at the head and tail of the first reach follow the same pattern and show more stability, the average standard deviation is about 0.07 m. The downstream water level dropped on 10 October due to more subtraction from the canal for pre-irrigation of wheat (preparation for winter crops). However, the upstream water level at the inlet accentuates a great discrepancy in October. This was due to the closure of most of the head regulators of the minor canals in that period since the rainfall satisfied the irrigation needs. This will be elaborated in Sub-section 5.3.4.

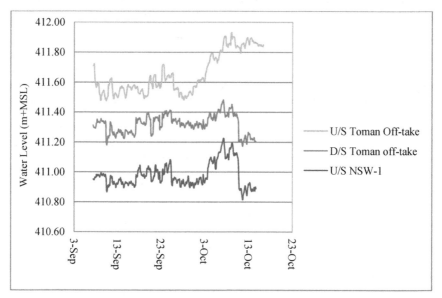

Figure 5.8. Water level at the offtake and tail of the first reach of Toman Minor Canal (7 September - 19 October 2012)

Figure 5.9 represents the water levels upstream and downstream of the control structures along the canal under consideration. The head was reasonable except at NS-2 where the head difference was reduced occasionally and there were indications water shortage in the third reach. Table 5.6 shows an increase in mean water level above FSL with reduces towards the downstream. According to the recorded water levels in 2012, at the period of low flow, the water level reduced below FSL in the third and fourth reach. The command also reduced towards the downstream.

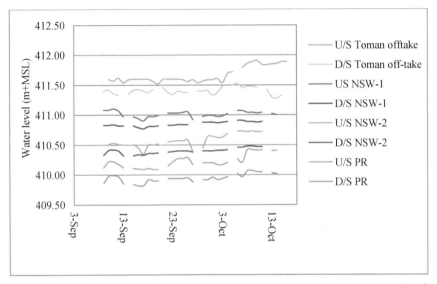

Figure 5.9. Water level at the head and tail of each reach along Toman Minor Canal (7 September - 19 October 2012)

Most of the time the farmers take the responsibility of adjusting the offtakes of the field canals. They tend to keep them fully open to divert the maximum possible without great consideration to the irrigation schedule. This results in surplus water as shown in Figure 5.10. In absence of reliable irrigation scheduling practices, the general tendency is inequity of water diverted to the system, over-toping and wastage of water.

Figure 5.10. Abundance of water as result of overtopping or farmer's mismanagement

Table 5.6. Water levels along Toman Minor Canal in 2012

Distance from offtake	Maximum level* (m)	Mean level* (m)	Minimum level* (m)	Ground level* (m)	FSL* (m)	Command at low level (m)	Command at mean level (m)	Min over FSL (m)	Mean over FSL (m)
0	411.5	411.31	411.19	410.65	410.8	0.63	0.75	0.48	0.6
1200 U/S	411.1	410.98	410.83	410.4	410.75	0.43	0.58	0.08	0.23
1200 D/S	410.91	410.85	410.77	410.4	410.61	0.37	0.45	0.16	0.24
3000 U/S	410.73	410.58	410.35	410.0	410.52	0.35	0.58	-0.17	0.06
3000 D/S	410.48	410.4	410.32	410.0	410.27	0.32	0.40	0.05	0.13
4700 U/S	410.52	410.25	409.92	409.82	410.19	0.10	0.43	-0.27	0.065
4700 D/S	410.07	409.95	409.82	409.82	409.94	0	0.13	-0.12	0.01

*all the levels are above MSL

5.3.3   Effect of change in bathymetry of the canal on the water level

Two cross-section surveys were performed at the beginning and end of the flood season for the two years. The aim was to determine the change in the bed profile and canal cross-sections and to assess the volume of deposition/scouring. The cross-section intervals were 200 m for the entire canals. It was performed during the last week of June and the first week of November in 2011 and in 2012 during the last week of June and the third week of September, before the planned clearance activities (desilting) by SGB for Zananda system started. It was found that the sedimentation after 20 September has no significant effect on the bed profile since the sediment concentration became very low. The results of the canal profile in 2011 and 2012 are presented in Figures 5.11 and 5.12. The cross-section surveys at the end of the flood season were not conducted for the second reach, since unexpected excavation had been detected in August 2011. The survey showed that the deposition along the first reach was 8,570 m³ (10,300 ton) and 5,900 m³ (7,100 ton) in 2011 and 2012 respectively.

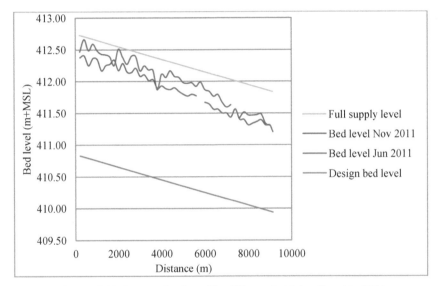

Figure 5.11. Longitudinal profile of Zananda Major Canal in 2011

The change in the canal profiles due to sedimentation during the flood season was identified and the longitudinal profile compared to the design. The clearance work is being performed improperly due to limited funds for maintenance. Thus, the deposition in the major canals and lower order canals kept on increasing over the years. It was found that the average rise in the bed level in Zananda Major Canal by the end of September 2012 was between 1.6 and 1.3 m (at the head and tail of the first reach) above the design level. More deposition has been detected at the head of the first reach. Figure 5.11 demonstrates that after the flood season of 2011 the bed level at the head of the first reach became just below the full supply level while at the end of September 2012 it became at the full supply level as is shown in Figure 5.12. The average rise of the bed during the study period was about 15 cm in 2011 and 11 cm in 2012. Figure 5.13 demonstrates the cross-section at 4.2 and 8.4 km from the offtake before and after the flood season in 2011.

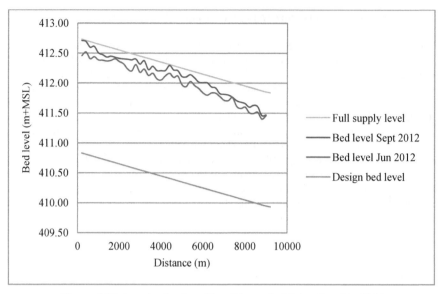

Figure 5.12. Longitudinal profile of Zananda Major Canal in 2012

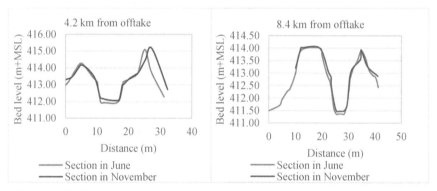

Figure 5.13. Comparison of initial and final cross-section during the flood season in 2011 at 4.2 km (left) and 8.4 km (right) from Zananda offtake

Figure 5.14 presents the result of the survey of Toman Minor Canal for 2012. It is shown that the bed level is far above the design level. The rise in the bed at the head of the canal is 1.2 m and it reduced to 0.9 m towards the downstream. Based on the surveys it was determined that the deposited amount of sediment is 1,320 m³ (1,580 ton), which represents 41% of the inflow sediment load of 3,880 ton till 20 September (the start of the survey).

The field outlet pipes became below the bed level of the minor canal and flow blockage problems due to piling up of sediment have been detected. Farmers overcome this problem by excavating deposited sediment in front of the field outlet pipes (FOP). Due to this water can find a pass from the minor to the FOP. Generally, the actual canal gradient is higher when compared with the design value. Figure 5.15 illustrates the bed level, ground level, mean water level and minimum water level along Toman Minor Canal between July and October in 2012.

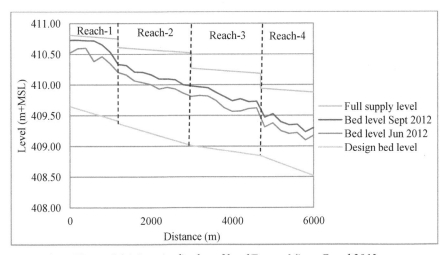

Figure 5.14. Longitudinal profile of Toman Minor Canal 2012

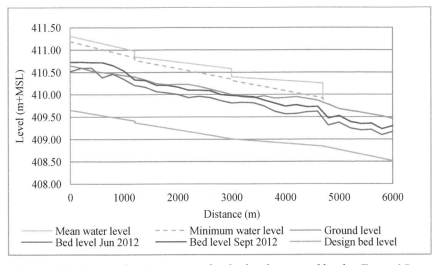

Figure 5.15. Mean and minimum water level related to ground level at Toman Minor Canal in 2012

The water levels were compared with the ground levels (at the initiation of the scheme from the design drawings) since there is no record of recent field levels. An increase in the ground level was expected as a result of the sediment that has been deposited in the fields during the last 90 years. Therefore, less command was expected. At low flow, the command reduced from 0.57 m at the offtake to 0.1 m downstream at the end of the third reach and to zero at the head of last reach as presented in Table 5.6. It is clear that at certain periods there is no head between upstream and downstream NS-2. In such cases farmers use pumps to withdraw the water from the minor to the field canals at the peak of the demand as shown in Figure 5.16.

5.3.4   Effect of the gate settings on the water level

The cross structures at the major and all the minor canals offtakes are moveable weirs. Table 5.3 shows the different types of structures and their locations at the canals under study. The distance between the bottom of the fixed beam and the crest of the weir is referred as gate opening, Figure 5.17 (a). The level of the bottom of the fixed beam represents the maximum height to which the weir crest can be raised. The upstream water level should not exceed the bottom of the weir which represents the maximum water level. The gate settings for most of the control structures have been measured on daily basis. It was found that in most of the time there is overflow above the fixed beam as shown in Figure 5.17 (b). This is related to the rise in bed level upstream of the control structures. The absence of systematic operation and lack of the adequate number of operators along the system leads the farmers take the responsibility of changing the gates setting randomly which reveal the instability of the water level.

Figure 5.16. Water shortage at the third reach of Toman Minor Canal

(a) Fixed beam at the movable weir   (b) Flow over the fixed beam

Figure 5.17. Different flow conditions of the movable weir

Table 5.7 presents the time percentage of exceedance of the fixed beam when compared with the total operation period during the study period in 2012. The percentage of exceedance recorded at Toman offtake was 80.6%, which denoted the highest value. Therefore, two calibration equations have been developed to cover both cases (normal and overflow conditions). For more detail, see Annex D. It was recognized that at cross strucure-1 (9.1 km fom the offtake) the water level never exceeded the beam level, since the position of the crest was set at its lowest position. In case of overflow, the discharge was estimated based on the calibrated structure and by assuming the fixed beam as sharp crested weir.

The gate setting of the movable weir was at the position of the weir crest with respect to the fixed beam. The measurements were conducted at the offtakes of the different minor canals along the first and second reaches and at the cross strucures in the major canal during the study period. The settings between 6 September and 16 October 2012 are shown in Figure 5.18 in order to find a correlation between the gate settings and the changes in the water levels. It was clear that the unsystematic change in the position of the crest level is the reason behind instability of the water levels as shown in Figure 5.6. The increase of the water level upstream of the offtake of Toman Minor Canal in October refered to the closure of most of the minor offtakes in that period. The changes would have to be made in sequence to achieve a quick stability of the system.

Table 5.7. Flow condition of different canals during the study period in 2012

Canal offtake structure	Operation (days)	Overflow (days)	Overflow (%)
Gimillia	37	29	78.4
G/Elhosh	62	13	21.0
Ballola	39	21	53.8
W/Elmahi	28	2	7.1
Toman	36	29	80.6
A/Gomri	29	7	24.1
Cross structure K12.5	87	32	36.8

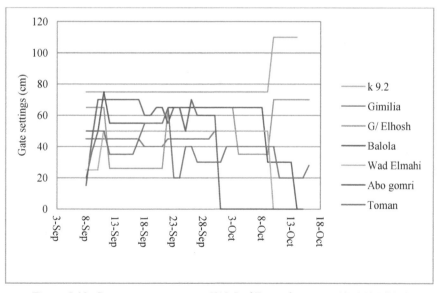

Figure 5.18. Gate settings upstream K12.5 of Zananda system 6/9-16/10/2012

Figure 5.19 illustrats the discharge and the change in the gate settings at the offtake of the major canal in 2012. It was recognized very low discharge (seepage) during the closure period and this amount was neglected. Although the sluice gates are appoximately fully opened during September and beginning of October, the discharge are low due to small head difference between upstream and downstream.

5.4   Flow measurements and calibration of measuring structures

Particularly during the last years the essential structural maintenance is not done. This type of maintenance is needed since it is necessary to allow the structures to be used for discharge measurement. The absence of this type of maintenance leads to improper measured discharge. Water can only be efficiently managed if it can be accurately measured. It is therefore important to calibrate the flow control structures. A series of measurements has been conducted on the water levels and the discharge in a rating section of known dimensions far from disturbances or distorting influence. The aim of flow measurement is to check the accuracy of the calibration equations of the control structures and to determine the mass balance at selected reaches.

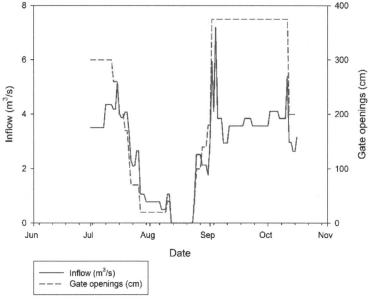

Figure 5.19. Changes in the setting of the crest level and discharge at K 12.5 at Zananda Major Canal

The flow velocities were measured by using a current-meter and the depth integrated method was carried out at the measurement locations. The flow velocity was measured in order to get the discharge passing through the canals. The canal cross-sections were divided into vertical sections depending upon the water surface width. For each section, the flow velocity was measured at 0.2 and 0.8 of the water depth (for water depth greater than 1.5 m) and at 0.6 of the water depth (for water depth less than 1.5 m). This is consistent with the U.S. Geological Survey (USGS) procedure. Then the average flow velocity has been computed. In most of the locations the discharge estimates based on the structure equations as applied by the operators showed over estimation of the discharge when compared with the measured data. The change in flow conditions as a result of sedimentation leads to a change in the discharge coefficients. Therefore, calibration has been conducted by using the measured data where it was needed. Calibration of the hydraulic structures has a prominent role in establishing the exact relationship between water stage and discharge for any given water level or gate opening (Kraatz and Mahajan, 1975). The actual releases at the inlets were derived from the calibration equation based on series flow measurements at different gate settings and flow conditions. A brief description of the stage discharge relations and coefficients for the flow conditions of the different structures has been elaborated in Annex D.

The water delivered by the Abu Ishreen canals was measured at selected canals by using a Parshall flume when the water level at Toman Minor Canal was low. The float method (surface float) was also adopted at high water levels at several field canals to estimate the discharge. The basic idea of the float method is to measure the time that it takes for the object to float a specified distance downstream. It measures the surface velocity, thus the mean velocity is obtained by using a correction factor. The correction factor was taken at 0.84 as a common used value (Oklahoma Water Resources Board, 2004). The measured discharge in the field canals varied between 0.02 and 0.12 m³/s. The clearance work between 14 and 18 September 2011 at Toman Minor Canal had a significant effect in improving the reliability of the water supply since the accumulated sediment upstream the field outlet pipes was removed. Figure 5.20 shows different techniques used for the flow measurements.

Figure 5.20. Flow measurement by using a current meter at the major and minor canals and a Parshall flume at the field canals

Tables 5.8 and 5.9 show the velocities measured at different locations along the canals. The average velocity at the offtake of Zananda Major Canal is 0.48 m/s and at Toman Minor Canal 0.36 m/s. Even the velocity reduced towards the downstream when more deposition was detected upstream. This confirms the fact that the fine sediment is not mainly dependent on the flow conditions. It was found that in most cases the velocity does not exceed the limit of 0.6 - 0.9 m/s for clay soils (Gismalla, 2009).

Table 5.8. Velocity along Zananda Major Canal (m/s)

Date/location	Offtake	At K 9.1	At K 12.5
20/7/2012	0.54	0.48	0.41
10/7/2012	0.51	0.41	-
16/10/2012	0.5	-	0.27
18/7/2011	0.48	0.41	0.39
13/9/2011	0.37	0.33	0.33
Average	0.48	0.38	0.35

5.4.1 Actual water release and crop water requirement

The term crop water requirement can be defined as almost equal to the amount of water required to compensate the evapotranspiration. In other words, the crop water requirement refers to the amount of water that needs to be supplied to meet the amount of water that is taken by evapotranspiration. It can be estimated based on the reference evapotranspiration (dependent on meteorological data such as temperature, relative humidity, solar radiation, sunshine and wind speed) and a crop factor, which is based on crop height, crop cover, leaf structure and root depth. The crop factor correspondingly depends on field capacity, wilting point, infiltration rate and bulk density. The bulk density of the clay changes appreciably with the moisture content (Ibrahim *et al.*, 1999).

Table 5.9. Velocity along Toman Minor Canal (m/s)

Date /Location	Offtake	2nd reach	3rd reach	4th reach
13/9/2012	0.33	0.37	0.31	0.17
27/9/2012	0.34	0.33	0.15	0.067
16/10/2012	0.42	0.36	0.18	0.11
17/10/2012	0.37	0.32	0.24	0.13
Average	0.37	0.35	0.22	0.12

The major crops in the study area are groundnut, sorghum, cotton, vegetables and wheat which is cultivated in winter. Based on the law of 2005, farmers have the right to plant what they need. The Sudan Cotton Company finances the farmers who grow cotton by supplying fertilizers and pesticides. It is also responsible for the land preparation and to sell the cotton. The income of the crop is shared between the company and the farmers. Due to the company policies with respect to payment (delay in payment) the farmers prefer to grow sorghum instead of cotton, which is also less costly. Thus cotton is not cultivated anymore in most of the command area. Figure 5.21 displays the cropping pattern in 2011 and 2012 in the area supplied by Zananda Major Canal as recorded by the SGB. It was found that in 2012 most of the cotton area was replaced by sorghum and vegetables. At the minor level, surveys were conducted in 2011 and 2012 to record the cropped area along Toman Minor Canal. Figure 5.22 shows the cropping pattern in 2011 and 2012 in the area supplied by Toman Minor Canal. The sorghum area represents the largest cultivated area. The crop intensity in the area supplied by Toman Minor Canal was 60% in 2011 and 48% in 2012.

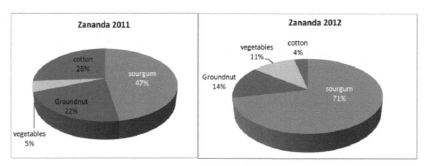

Figure 5.21. Cropping pattern in the area supplied by Zananda Major Canal

The cropped areas supplied by Zananda Major Canal were about 3800 ha in 2011 and 4680 ha in 2012 as shown in Table 5.10. More water was released in 2011 between mid-July and 10 August as shown in Figure 5.23. This period coincides with the period of high sediment concentration. As a result, more sediment was entering the scheme in that period. However, in 2012 due to heavy rainfall during the flood season the amount of water supply to the scheme was reduced. Hence, the crop water requirement was more than the released amount since no supplementary irrigation was needed. In contrast in 2011 the rainfall occurred in a short period. The delivery performance ratio, which is the ratio between the actual discharge and the target discharge (crop water requirement) has been tested. It was found that the delivery performance ratio was 1.15 in 2011 which indicates an oversupply.

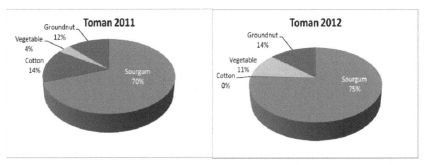

Figure 5.22. Cropping pattern in the area supplied by Toman Minor Canal

Table 5.10. Cropped area (ha) in the study area

Crop	Zananda Major Canal		Toman Minor Canal	
	2011	2012	2011	2012
Sorghum	1820	3340	320	271
Cotton	966	163	65	0
Vegetable	177	528	20	8
Groundnut	840	654	56	95
Total cropped area	3803	4684	460	374
Command area	8520	8520	772	772
Fallow	4717	3836	312	398
Crop intensity %	45	55	60	48

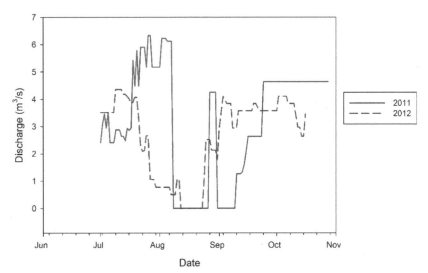

Figure 5.23. Water released at the offtake of Zananda Major Canal

The crop water requirements including field losses were computed for the different periods based on Farbrother's (1974) intensive study in Gezira Scheme in the early seventies. The results of his study are presented in Table 5.11. Adam (2005) stated that the recent studies confirmed that Farbrother's factors are still valid and can be applied in Gezira.

Table 5.11. Crop water requirement in m³/ha/day

Month	Period	Cotton	Groundnuts	Sorghum	Gardens
June	1-10		42.4	190.5	63.3
	11-20		41.9	39.5	63.3
	21-31		43.1	40.2	58.6
August	1-10	71.4	46.2	47.6	54.5
	11-20	32.4	50.7	61.2	51.9
	21-31	32.9	60.0	72.6	52.9
September	1-10	38.8	68.8	76.2	54.5
	11-20	46.2	75.2	76.0	55.2
	21-31	58.6	75.9	76.0	55.2
October	1-10	67.4	72.9	70.7	54.5
	11-20	73.8	67.9	58.6	52.9
	21-31	76.9	56.9	41.7	51.2
Sowing date		1 - 10 Aug	21 - 30 Jun	1 - 10 Jul	

(Source: Farbrother, 1974)

The Farbrother's factors, which have been determined by field measurements cover all losses below the FOP level. The canal conveyance and distribution efficiency were estimated at about 93% and the overall efficiency at 70% when the canals are in an acceptable condition, which is one of the highest in the world (Plusquellec, 1990). This referred to the impermeable clay soils in the scheme and the role of storage in the minor canals. The efficiency is expected to have been reduced during the last years, since the canals were in a worse condition due to the absence of systematic maintenance. However, since there are no data available on the overall efficiency of the Gezira irrigation system, the ratio between crop water requirement (CWR) including field losses (by applying Farbrother's factors) and the actual measured flow have been computed. This ratio was computed to evaluate the beneficial use of the water delivered to the system. The average percentage of this ratio was 67.7% that is an indicator of improper water management in the system as shown in Table 5.12.

Table 5.12.Crop water requirement (CWR) and actual released in 2011

Period	CWR (Mm³)	Actual flow (Mm³)	% of the ratio CWR/actual released
September III	2.68	3.49	76.9
October I	2.65	4.01	66.0
October II	2.44	4.01	61.0
October III	2.28	3.41	66.7
Average			67.7

The losses in the canals are mainly due to evaporation. The average monthly evaporation in July, August, September and October is 244, 206, 206 and 204 mm respectively (Adam, 2005). The evaporation in the Zananda system has been computed. The lengths of the major canal and all minor canals, supplied by the major canal, are about 17 and 66 km respectively. When the canals were supplied at their full capacity the surface width was about 15 and 8.4 m for the major and minor canals respectively. It was found that the evaporation losses in the Zananda system was about 0.02 Mm^3 during the flood season (4 months). However, the evaporation losses were rather small and represented about 0.03% of the supply when the canals were operated with their full capacity (the release is 58.4 Mm^3) during that period.

The water released (considering the losses) was compared with the actual need by the crops as shown in Figures 5.24, 5.25 and 5.26. At the scheme level the released water was extremely high in comparison with the crop water requirement. The Sudan Gezira Board referred that the actual cropped area may have been larger than what was reported. In addition the stage discharge relationships of the sluice gates at the offtakes of the main canals that are still in use overestimate the discharge and need calibration. The peak demand at the beginning of July is due to the first irrigation of sorghum to eliminate the deficit in the soil moisture below the wilting point. In 2012 the reason behind the higher crop water requirement in July I was the increase in the cropped area of sorghum by 71% of the total cropped area, when comparing Figure 5.25 with Figure 5.26. The crop water requirement in July 2012 for Zananda Major Canal was higher than the canal capacity 5.5 m^3/s (4.75 Mm^3/10 days). Moreover, the crop water requirement also exceeded the full capacity 0.46 m^3/s (0.4 Mm^3/10 days) in Toman Minor Canal in 2011 and 2012. HR Wallingford (1990a) proposed to reduce the amount of the first irrigation by means of pre-irrigation, which can be extended to more than a week. The maximum authorized discharge based on the duty of 71.4 m^3/ha/day was 3.1 m^3/s in 2011 and 3.9 m^3/s in 2012. This indicated that in 2011 the inflow also exceeded the authorized one in most cases (over supply) when excluding the rainfall period.

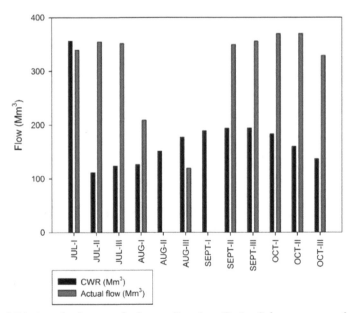

Figure 5.24. Actual release at the Sennar Dam into Gezira Scheme compared with the crop water requirement for 2011

The heavy rainfall events were measured during July and August by using a simple rain gauge that was installed in the open space near the Zananda offtake. The results are summarized in Table C.6. The total rainfall during July and August coincided with that recorded at Wad Madani station by the meteorological corporation as shown in the Tables 3.1 and 3.2. The estimated water supplied by the rainfall during July and August was related to the CWR to assess how the effective rainfall (80% of actual rainfall) satisfied the demand. Table 5.13 shows that although the rainfall satisfied the demand by 77% in August 2011, there was more water delivered to the system (about 5 Mm^3). However, in August 2012 the rainfall and delivered water was less than the demand, but there was no complaint about water shortage since the rainfall was distributed uniformly during the month. The results indicate water mismanagement and oversupply during most of the time in the period of high sediment concentration.

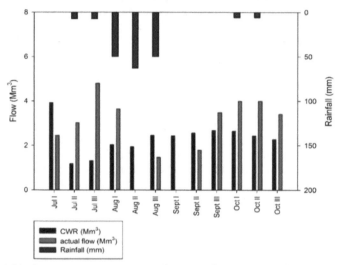

Figure 5.25. Crop water requirement and water release to Zananda Major Canal in 2011

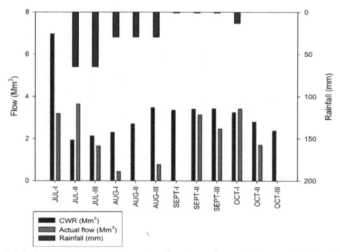

Figure 5.26. Crop water requirement and water release to Zananda Major Canal in 2012

Table 5.13. Summary of actual demand and rainfall in the Zananda Major Canal commanded area during July and August

	2011		2012	
	July	August	July	August
CWR (Mm³)	6.44	6.45	11.03	8.46
Effective rainfall (mm)	12	130	73	106
Cropped area (ha)	2840	3800	4520	4680
Water supply by rainfall (Mm³)	0.34	4.96	3.29	4.98
% of rainfall satisfy the demand	5.2	76.9	29.8	58.9

At the minor level the water supply to the minor is not far from the crop water requirement as is shown in the Figures 5.27 and 5.28. In 2012 the closure of the minor occurred most of the time during the flood season, since the rainfall satisfied the demand. This prevented the minor from more sediment deposition.

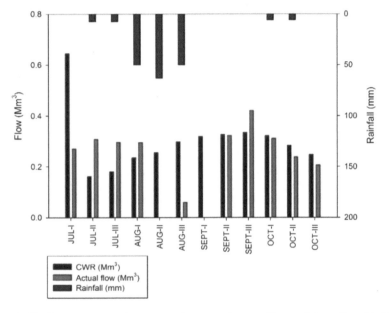

Figure 5.27. Crop water requirement and water release to Toman Minor Canal in 2011

5.4.2 Operation and irrigation schedule of Toman Minor Canal

The opening of the field canals (Abu Ishreens) at the 1st, 2nd and 3rd reaches of Toman Minor Canal were recorded on a daily basis during September 2012. The data are summarized in Table C.8 (Annex C). The operation of the Abu Ishreen at the first reach during September and October 2011 was recorded on a daily basis. It was recognized there is no regular irrigation schedule followed. Most of Abu Ishreen canals were opened most of the time. The monitoring of the field canals was reported in the morning. However, the reason behind not reporting on part of the Abu Ishreen canals, may be due to the fact that farmers opened the FOP for different duration later during the day.

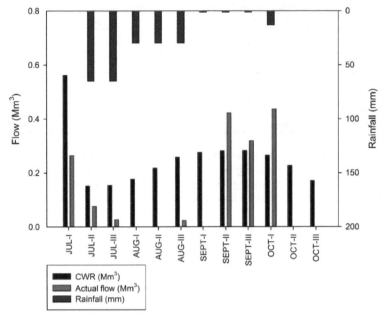

Figure 5.28. Crop water requirement and water release to Toman Minor Canal in 2012

Occasionally, excess water delivered to the canals during the flood season was resulting in breakage of most of the minor canals, especially in July and August 2011. This can also be referred to heavy rainfall when most of the offtakes were closed. The absence of the role of the water users association let the farmers take the responsibility for the operation of the Abu Ishreen canals, without any coordination among them. This created inequity in the water distribution. The stress in water at Toman Minor Canal was found in reach-3 since there was less command as a result of raising of the bed level due to sedimentation as mentioned in Sub-section 5.3.3.

It was found that at low flow the Nimra (37.8 ha) needs more than two weeks to be irrigated when the canal operates for 24 hours since the Abu Ishreen discharge varies between 0.02 and 0.12 m³/s. In the past when the canals were in a good condition, the Nimra needed 7 days to be irrigated when the field canals were operated for 12 hours at full capacity of 0.115 m³/s.

5.5   Sediment analysis

5.5.1   Suspended sediment concentration

Suspended sediment concentrations were measured at 13 different locations as shown in Figure 5.2. Samples were taken on a daily basis during the flood season in 2011. From the end of September the samples were taken every two days, since the variation in concentration was not significant. In 2012 the samples were taken on a two days basis and with 3 days interval from end of September. Samples were collected from the middle, downstream of the hydraulic structures where the water was well mixed. The accuracy of the measured sediment concentration depends on the accuracy of the field sampling and laboratory techniques. The dip sampling technique was chosen among other techniques, while fine sediment is more uniformly dispersed through the section

(Scarlatos and Lin, 1997, Edward and Glysson, 1999) and based on a study conducted at Gezira Main Canal by using different sampling techniques: dip-sampling and depth integrated sampling by using a US D-74 sampler. The study found close correlation and good fitting between the two methods with a correlation coefficient of 0.97 as shown in Figure 5.29. The dip sampling technique was recommended to be used for sediment sampling in Gezira Scheme (HR Wallingford, 1990a). The sediment sampler has been manufactured to collect samples from middle of the water column. Figure 5.30 presents the sediment sampler that was used. The long handle on the sampler allows access from a discrete location.

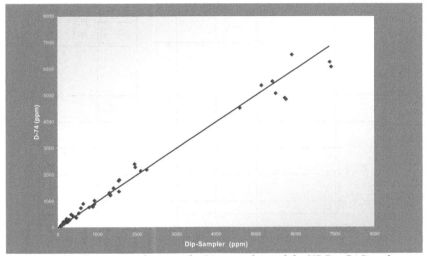

Figure 5.29. Comparison between the Dip-sampler and the US D – 74 Sampler at Gezira Main Canal downstream of km 108 (Hussin, 2006)

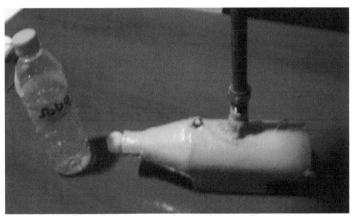

Figure 5.30. Sediment sampler for dip sampling technique

The Figures 5.31 and 5.32 present the sediment concentration in Zananda and Toman canals respectively. The peak concentration occurs at the end of July or the beginning of August. It was about 14,000 ppm at Zananda Major Canal offtake. At Toman Minor Canal offtake it was about 11,600 ppm in 2011 and 10,000 ppm in 2012.

The period of gate clousure at the offtake of the canals is also presented in the garph since no supplimentary irrigation was needed. This period prevented the system of more inflowing sediment load. At the beginning of the flood season the sediment concentration increased drastically. The period of high sediment concentration was found from mid-July till the end of the first week of August, which coincides with the findings of HR Wallingford (1990a). Then the sediment started to reduce gradually to less than 1000 ppm in October and reached less than 200 ppm in November. The rising limb of the sediment hydrograph is very sharp, while the falling limb is gradually.

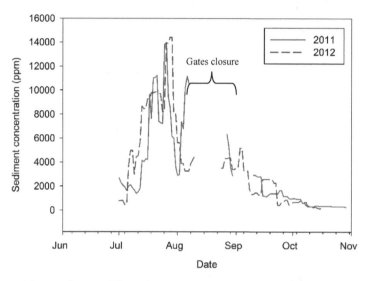

Figure 5.31. Distribution of the sediment concentration at the offtake of Zananda Major Canal during the flood season in 2011 and 2012

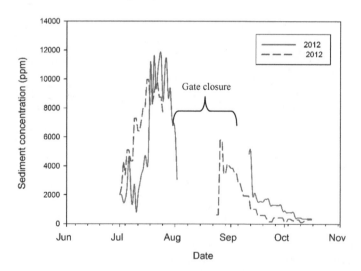

Figure 5.32. Distribution of the sediment concentration at the offtake of Toman Minor Canal during the flood season in 2011 and 2012

5.5.2 Settling velocity in the study area

The settling velocity is varying temporally and spatially with the concentration and is considered as one of the most important parameters in cohesive sediment transport. The physical sediment properties of sediment such as size, shape and specific weight are expressed as fall velocity. HR Wallingford (1990a) measured the settling velocities at different concentrations of suspended sediment in the pilot area of the study (Zananda and Toman canals) for five sediment factions by using an Owen tube. The results of the study are plotted in Figure 5.33.

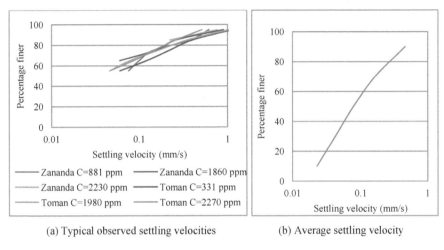

(a) Typical observed settling velocities   (b) Average settling velocity

Figure 5.33. Observed settling velocities in the study area and in Gezira Scheme
(HR Wallingford, 1990a)

The correlation between measured settling velocity of the mean concentration and the mean settling velocity for the 75% sediment fraction is presented in Figure 5.34. The settling velocity increases gradually with the increase in sediment concentration as shown in the following equation:

$$w_s = 2 \times 10^{-7} C^{0.8}$$ (5.1)

Where:
w_s = settling velocity (m/s)
C   = concentration of sediment (ppm)

Figure 5.35 shows the concentration at 0, 30, 60 and 90 cm depths at the offtake of Zananda Major Canal on 18 July 2011 (left graph) and at 0, 35 and 60 cm water depths on 13 September 2011 (right graph). Although the number of the data is limilted, the sediment profiles demonstrate that the variation in concentration with depth is very small since the sediment is very fine and the difference in concentration between mid-depth and the near the bed is not significant.

Figure 5.36 shows a family of curves obtained by plotting the Rouse Equation (4.12) for different values of the exponent z_i (Sub-section 4.4.2). The vertical distribution of the sediment in Gezira Scheme has been plotted in the graphical representation of the Rouse Equation. The wash load in the graph refers to clay and silt.

For low values of z_1, which is the case in this study, the sediment distribution is nearly uniform over the water column (Simons and Senturk, 1992). This was also found when the sediment samples were measured in different water columns, in order to develop the vertical suspended sediment profile during the study period.

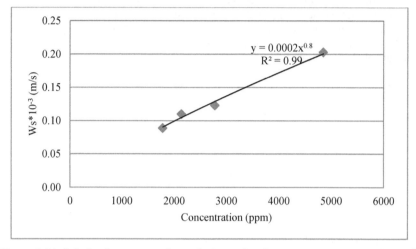

Figure 5.34. Relation between settling velocity and sediment concentration in Zananda Major Canal

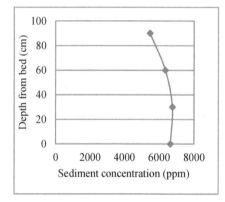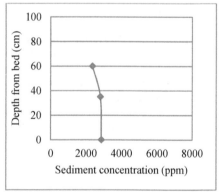

Figure 5.35. Sediment profile at Zananda offtake on 18 July and 13 September 2011

5.5.3 Sediment load and mass balance study

The large flow delivered to Gezira Main Canal (between July and October on average 17 and 14.5 Mm³/day in 2011 and 2012 respectively) with high velocity about 0.7 m/s reduced the amount of deposition. Figure 5.37 shows the concentration at the offtake of Gezira Main Canal downstream of the Sennar Dam and the concentration at the offtake of Zananda Major Canal at 57 km from the Sennar dam. The graph indicates less deposition (especially in 2011 the change of concentration is not that much) in the main canal. In 2011 the sediment concentration at the offtake of Zananda Main Canal was measured on a daily basis, whereas in 2012 it was measured every two days and the values were considered as average for two days.

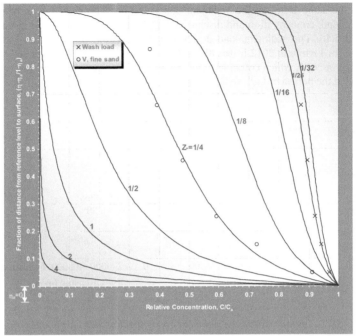

Figure 5.36. Representation of the Rouse Equation for the vertical distribution of suspended sediment in Gezira Main Canal downstream km. 108 (Hussin, 2006)

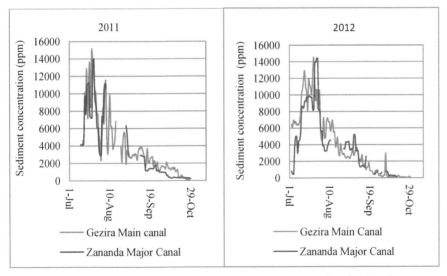

Figure 5.37. Concentration at the offtake of Gezira Main Canal and Zananda Major Canal at 57 km from the Sennar Dam during the flood period in 2011 and 2012

Figure 5.38 demonstrates a high correlation coefficient between the sediment concentration and sediment load by using the data collected in 2011. The sediment load is more dependent on the sediment concentration rather than on the discharge and follows the same trend as the sediment concentration. The correlations between the sediment load, concentration and discharge in 2011 and 2012 are plotted in the Figures 5.39 and 5.40. The sediment load during the flood season in 2011 was 118×10^3 ton (98.4×10^3 m^3) while in 2012 it was 88.1×10^3 ton (73.4×10^3 m^3). A higher sediment load took place in 2011, which coincides with the excess discharge above the full capacity of the canal at the period of high sediment concentration.

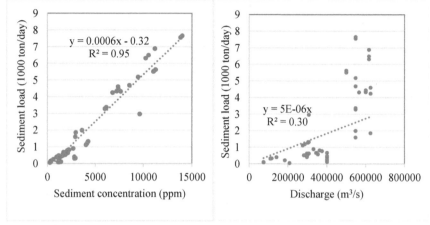

Figure 5.38. Correlation between sediment concentration, discharge and sediment load

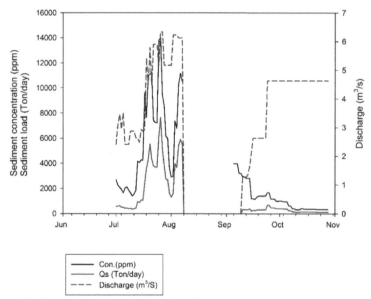

Figure 5.39. Relation between sediment load, concentration and discharge at the head of Zananda Major Canal between July and October 2011

Table 5.14 summarizes the supply of water and sediment to the study area. HR Wallingford (1990a) reported that the average sediment supply was 4.6 ton/ha in 1988 and 1989. This means an increase of more than a ton per hectare during the last years.

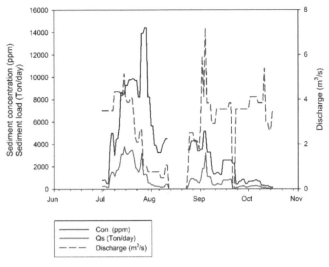

Figure 5.40. Relation between sediment load, concentration and discharge at the head of Zananda Major Canal between July and October 2012

Table 5.14. Summary of sediment and water release to Zananda Major Canal commanded area

Canal	2011	2012
Gross area under command (ha)	8,520	8,520
Water supply (Mm³)	28.1	24.2
Total sediment load (ton)	118,000	88,100
Volume of water supply/ha (m³)	3,300	2,840
Mass of sediment supply/ha (ton)	5.9	5.7
Mass deposited (ton) at 1st reach according to survey	10,300	7,080
Average bed rise (cm)	15	11

Sediment budget of the first reach of Zananda Major Canal

In order to study the hydraulic and sediment transport functions of the canals and the interdependence between hydraulic, morphologic and sediment parameters, mass balances were conducted between July and October in 2012. For the development of a conceptual model it is useful to establish a rough sediment mass balance. The total discharge and average sediment concentration at the inlet and outlet of the first reach were measured and are summarized on a weekly bases as presented in Tables 5.15 and 5.16.

The sediment load and deposition were computed and are presented in Table 5.17. It was observed that during week 7 and 8, the offtake was closed because of rainfall and the seepage flow through the gate was ignored. More deposition was detected during the third week. The deposition in the first reach was 12.8×10^3 ton in 2012. About 60% of the deposition occurred in July. The deposition till week 12 was 12.0×10^3 ton while in 2012 it was 10.4×10^3 ton. It was recognized that more deposition was detected by the mass balance. This may be related to illegal withdrawal from the major canal (at 2.8 km and 10 km from the offtake) that was observed during the study period.

Table 5.15. Total discharge in Mm^3 per week

Week	Zananda offtake	Gimillia	G/Elhosh	Bellola	W/ Elmahi	K 9.1
1	2.12	0.17	0.13	0.15	0.16	0.81
2	2.53	0.11	0.19	0.18	0.23	1.71
3	2.46	0.13	0.21	0.10	0.03	1.90
4	1.21	0.01	0.06	0.03	0.00	1.01
5	0.49	0.00	0.00	0.00	0.00	0.40
6	0.45	0.00	0.00	0.00	0.00	0.36
7	0.50	0.00	0.00	0.00	0.00	0.41
8	0.50	0.00	0.00	0.00	0.00	0.41
9	1.40	0.09	0.05	0.08	0.10	1.04
10	2.74	0.20	0.12	0.13	0.12	2.04
11	2.05	0.16	0.13	0.18	0.28	1.22
12	2.13	0.21	0.09	0.20	0.28	1.39
13	2.16	0.21	0.12	0.21	0.29	1.21
14	2.39	0.36	0.13	0.26	0.30	1.23
15	2.31	0.26	0.19	0.15	0.16	1.58

Table 5.16. Average sediment concentration (ppm)

Week	Zananda offtake	Gemilia	G/Elhosh	Ballola	W/ Elmahi	K 9.1
1	2230	2610	4040	2620	5650	978
2	6350	3580	4750	7240	7640	5840
3	9630	7180	9130	9630	5870	8420
4	11200	8770	9780	0	0	12100
5	5910	0	0	0	0	6880
6	3260	4110	0	0	5050	3780
7	0	0	0	0	0	0
8	0	0	0	0	0	0
9	3900	4810	4680	4510	4510	2500
10	3920	3020	3110	3020	3310	3710
11	1670	1720	2070	1810	1910	1400
12	2320	2140	1100	611	564	2200
13	680	814	684	479	557	659
14	721	235	843	153	336	873
15	436	197	285	40	315	362

A high sediment concentration was detected at the outlet of the canal under study. It exceeded the concentration at the inlet. The reason was that the control structures are in a bad condition. Farmers use sacks filled with soil to adjust the water released downstream as shown in Figure 5.41. This increases the concentration as well.

Sediment budget of the first reach of Toman Minor Canal

The first reach has 5 Abu Ishreen canals in addition to two field canals (called Nackosi) in the other direction to irrigate a Nimra that must be irrigated from other minor canal as shown in Figure 5.3 (referred to by n2 and n4). One out of the two Nackosi was in operation. The sediment budget of the first reach of Toman Minor Canal has been determined for a week in 2011 as depicted in Figure 5.42. Sediment samples have been collected on a daily basis and the irrigation schedule during that week has been reported. The sediment concentration and flow were measured at the field outlet pipes on a daily basis. It was found that the deposition in the first reach of Toman Minor Canal was about 25% of the inflowing sediment.

Table 5.17. Sediment load and deposition at the first reach in 1000 ton

Week	Zananda Offtake	Gemilia	G/Elhosh	Ballola	W/ Elmahi	K 9.1	Deposition
1	4.74	0.44	0.52	0.4	0.92	0.8	1.66
2	16.1	0.38	0.91	1.3	1.75	10	1.74
3	23.7	0.9	1.9	0.97	0.17	16	3.73
4	13.5	0.12	0.61	0	0	12.2	0.56
5	2.91	0	0	0	0	2.76	0.15
6	1.46	0	0	0	0	1.35	0.11
7	0	0	0	0	0	0	0
8	0	0	0	0	0	0	0
9	5.45	0.45	0.22	0.36	0.48	2.61	1.34
10	10.7	0.6	0.39	0.4	0.39	7.59	1.38
11	3.42	0.27	0.27	0.32	0.53	1.7	0.32
12	4.95	0.46	0.1	0.12	0.16	3.06	1.05
13	1.47	0.17	0.08	0.1	0.16	0.8	0.16
14	1.72	0.09	0.11	0.04	0.1	1.07	0.31
15	1.01	0.05	0.05	0.01	0.05	0.57	0.27
Total	91.1	3.93	5.16	4.02	4.71	60.5	12.8

Figure 5.41. Sacks and grass used by farmers to control the water delivered to Gemoia Minor Canal

The concentrations were measured downstream of the control structures at the offtakes, NS-1 (at 1.2 km), NS-2 (at 3.0 km) and PR (at 4.7 km) as presented in Table 5.18. The location of the control structures in the Toman Minor Canal is shown in Figure 5.3. In general the concentration decreased towards the downstream. The high reduction percentage of the concentration in the last reach indicates high sediment deposition or that more sediment passes to the fields.

5.5.4   Comparison of the sediment load with historical data

The data of the water released between July and October from the Sennar Dam during the previous 16 years illustrate an increase in the water supply to the scheme through the Gezira Main Canal. The Gezira Main Canal operates with more than its full capacity (14.5 Mm³/day) for most of the time. Figure 5.43 indicates a slight increase in the water delivery over the years and a gradually increase in the sediment inflow to the scheme. The increase in the sediment load is due to increase in concentration over time as mentioned in Sub-section 3.3. Figure 5.44 shows that the average sediment concentration varies significantly with time. The average monthly sediment concentration can be derived from the data that were collected by the Hydraulic Research Centre of the Ministry of Water Resources and Electricity between 1996 and 2009. The reason behind low the concentration in July between 2007 and 2009 is not clear. The sediment loads released to Gezira Main Canal were 6.85×10^6 ton in 2011 and 6.17×10^6 ton in 2012. Ali (2014) confirmed the increase in the sediment over the years and reported more erosion in the upper Blue Nile River Basin.

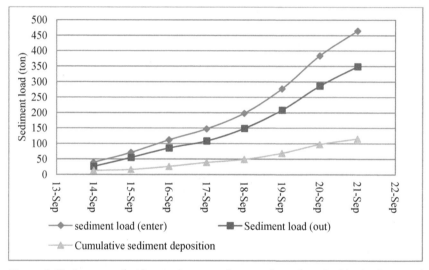

Figure 5.42. Amounts of sediment that enter, leave and are deposited in the first reach of Toman Minor Canal in 2011

The water supply, sediment concentration and sediment load supplied at Zananda canal in 2011 and 2012 has been compared with the historical data collected previously between July and October as shown in the Figures 5.45 and 5.46. The aim of this comparison was to relate these two years with the previous years. It is recognized that in 2011 the sediment load was high during the period of peak sediment concentration between July and early August and that more water was delivered to the system during that period. The increase in the water supply between week 2 and week 6 was about 33% more than the average value. Inevitably, this led to an increase in the sediment load with about 139% above the average of the same period. However, in 2000 more water was delivered to the minor after week 9. However, this had no effect on the sdiment since the concentration was low. This confirmed that the sediment load is predominantly dependent on the sediment concentration than on the discharge.

Table 5.18. Sediment concentration along Toman Minor Canal in ppm

Date	Offtake	NS-1	NS-2	PR	Reduction (%)
05/07/2012	5110	1440	3110	1040	79.6
07/07/2012	4330	4200	2740	1480	65.8
09/07/2012	7300	4540	5930	1690	76.8
18/07/2012	8190	7930	6510	6130	25.2
21/07/2012	9740	8100	8020	8880	8.8
27/07/2012	7670	6970	4710	3130	59.2
27/08/2012	4630	1340	1440	698	84.9
28/08/2012	3460	2690	1580	808	76.6
01/09/2012	4180	2760	1680	1170	72.0
03/09/2012	3540	3050	1830	1480	58.2
19/07/2011	8390	7690	----	592	92.9
13/09/2011	1890	769	868	277	85.3
17/09/2011	1710	875	870	661	61.3
18/09/2011	1100	737	523	262	76.2
04/10/2011	661	523	246	222	66.4

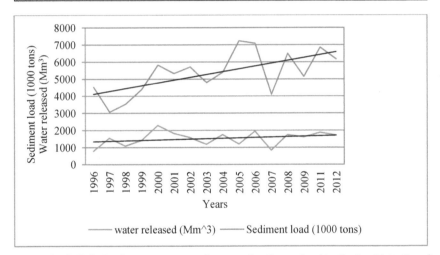

Figure 5.43. Relation between water release and sediment load in Gezira Main Canal between July and October

5.6   Laboratory analyses

Sediment concentration, specific weight of sediment, sieve analysis and hydrometric analysis for sediment sizes were accomplished. Physico-chemical analysis, Atterberg limits test ad water column test were performed to study the properties of the sediment. Details of these analyses will be presented in the following sections.

5.6.1 Suspended sediment concentration

The two most commonly used methods for measuring the sediment concentration are evaporation and filtration. The filtration is used for coarser grain sizes and lower concentration. The evaporation technique was used since it is recommended for finer materials according to the techniques of water resources investigation of the U.S. Geological Survey (Guy, 1969). The method consists of allowing the sediment to settle to the bottom of the sample bottle, decanting the supernatant liquid, washing the sediment into an evaporation dish and drying it in an oven.

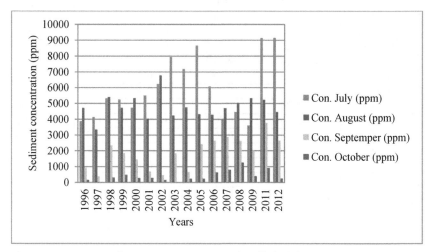

Figure 5.44. Sediment concentration at the offtake of Gezira Main Canal

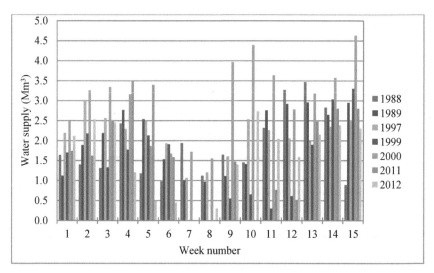

Figure 5.45. Comparison of water release to Zananda Major Canal for several years

5.6.2   Grain size distribution

The fundamental property of sediment is the particle size. It is essential to measure the grain size distribution for sediment classification. The grain size distributions were obtained for both bed material and suspended sediment transport. The sieve analysis was used for grain sizes of more than 0.063 mm (limit between sand and silt). The hydrometer test was conducted for fine materials (silt and clay). Dispersing agent, a solution of sodium hexametaphosphate in distilled or demineralized water, at the rate of 40 g of sodium hexametaphosphate per litre of solution was used in the hydrometer test to separate collides and to remove the organic matter.

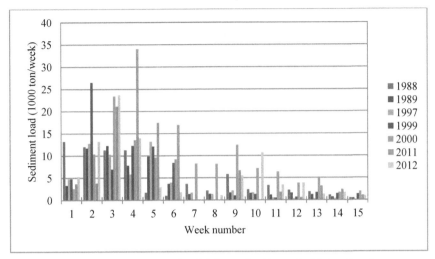

Figure 5.46. Sediment load delivered to Zananda Major Canal

Bed material samples have been taken to obtain the sediment grain size distribution of the deposited sediment and the specific weight of the sediment. The bed samples were collected by using the Van Veen bottom grab at the offtake of Zananda and Toman canals and at the tail of the Toman Minor Canal. It was found that the average specific weight of the sediment is 2.74. At the beginning of the flood season the fine material is washed from the river basin and transported through the irrigation canals. The analysis of bed material at Toman offtake provided about 85% silt, 9% clay and 6% fine sand while the sediment at the last reach consists of 70% silt and 30% clay as shown in the Figures 5.47 and 5.48.

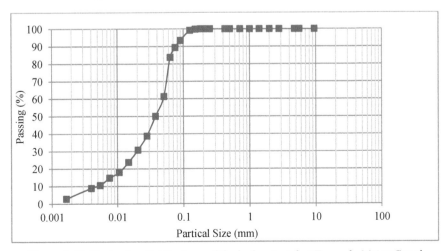

Figure 5.47. Grain size distributions of the bed material at Zananda Major Canal

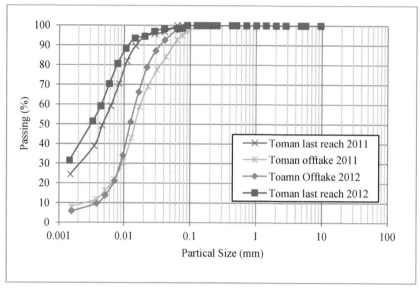

Figure 5.48. Grain size distribution of the bed material at Toman Minor Canal

The grain size distribution of the bed material is summarized in Table 5.19. The dry sediment materials that remained from the sediment concentration tests have been collected and kept for each measured location separately throughout the season. The samples at the end of the season can be considered as representative samples for the suspended sediment for the whole season. The results of the sieve and hydrometer tests for the suspended material show that the sediment contains about 70% silt and 30% clay as illustrated in Figure 5.49. It was recognized that no sand fraction has been detected. Based on the grain size distribution the sediment is considered as cohesive sediment since it has the identified cohesive size (Mehta *et al.*, 2013).

Table 5.19. Grain size distribution of bed materials for different samples

Location	Year	Fine sand (%)	Silt (%)	Clay (%)	Classification according to Shepard (1954)
Zananda offtake	2011	25	71	4	Sandy silt
Toman offtake	2011	10	80	10	Silt
	2012	2	90	8	Silt
Toman last reach	2011	0	70	30	Clayey silt
	2012	1	69	30	Clayey silt

At higher clay contents (4 - 10%), the structural framework of the sediment changes from a sand grain framework to a clay mineral framework, which corresponds to a shift from a non-cohesive to a cohesive sediment (van Ledden *et al.*, 2004; Winterwerp and van Kesteren, 2004). The sediment has been classified according to Shepard's diagram based on the proportions of sand, silt and clay particles. This classification system is the basis of the sediment distribution map (Shepard, 1954). The sediment became finer towards the downstream, since the coarser materials deposited first upstream. The canals under study were located upstream of Gezira Main Canal so no fine sand was expected downstream.

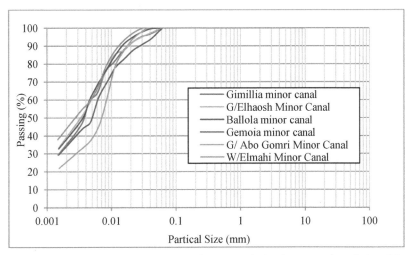

Figure 5.49. Grain size distribution of the suspended sediment at the offtake of the different minor canals

5.6.3   Atterberg limits test

Atterberg limits test provides the basis for a very simple but effective classification system for unconsolidated sediments (Keller and Bennett, 1973). The aim of this test is to distinguish between cohesive and non-cohesive sediment and to determine the mechanical behaviour of cohesive sediment. Liquid and plastic limits yield the water content of cohesive sediment samples. The difference in water content between the liquid and plastic limit is called the plasticity index (PI = LL-PL). The plastic limit is the water content at which the sediment sample can be rolled into threads 3 mm in diameter without breaking into pieces. However, the liquid limit is measured by using two half of sediment cake that will flow together for a distance of 1.25 cm along the bottom of a groove separating the two halves when the cup containing the cake is dropped 25 times for a distance of 1 cm at the rate of two drops per second.

The sediment has been classified according to the existing organic and plasticity level by using a plasticity plot. Table 5.20 shows the results of the analysis. Another application of Atterberg limits is when PI is related to the clay fraction in the so called activity-plot. From this test it was verified that the sediment is cohesive, since it has medium to high plasticity.

Table 5.20 .Atterbarg limit test for Toman Minor Canal soil

Location	Liquid limit (LL) (%)	Plastic limit (PL) (%)	Plasticity index (PI) (%)	Classification
Toman Offtake	45	31	14	Silt and organic clay, medium plasticity
Toman last reach	69	38	31	Silt and organic clay, high plasticity

Atterberg limits provide additional information on the plastic behaviour of the sediment as a whole and on the clay fraction of the sediment. When PI increases, the cohesion increases (Grim, 1962; Whitlow, 1995). However, the utility of these measures for predicting the erodibility of superficial sediment is questionable (Partheniades, 2007). The mechanical soil properties can be linked to erosive properties of cohesive sediment. Smerdon and Beasley (1959) correlated the critical shear stress of erosion to the plasticity index of bed material according to the following equation:

$$\tau_c = 0.163(PI)^{0.84} \qquad\qquad\qquad (5.2)$$

Where:

τ_c = erosion critical shear stress (N/m^2)

PI   = plasticity index (%)

According to the mechanical properties of the sediment at the offtake of Toman Minor Canal τ_c = 0.031 psf (1.48 N/m^2) and for the last reach it is 0.061 psf (2.92 N/m^2). From these values it was found that the critical shear stress exceeded the bed shear stress. This confirms the fact that erosion does not exist in the Gezira irrigation canals. The critical shear stress may also vary widely for soils with similar characteristic parameters by orders of magnitude (Partheniades, 2009). Therefore these values were not considered in this study. The mechanical parameters are secondary indices not necessarily representing the detailed internal structure of cohesive soil. However, chemical tests have been carried out to study the inter-particle physico-chemical ponds.

5.6.4 Analysis of physico-chemical properties of the sediment (chemical analysis tests)

The physio-chemical properties of the sediment will be discussed based on the results of the laboratory analysis for two samples from the offtake of Toman Minor Canal as presented in Table 5.21

Table 5.21. Laboratory determinations used to evaluate the sediment properties

Determinations	Sample-1	Sample-2
SP %	76.6	111.1
PH	7.57	7.57
EC (dS/m)	0.57	0.54
Ca+Mg (mmol+/l)	4	4.2
Na (mmol+/l)	1.6	1.2
K (mmol+/l)	0.13	0.13
SAR	1.1	0.8
Exh. Na (mmol+/100g)	1.05	0.97
ESP %	3.3	2.9

The saturation percentage (SP) is an indication of soil water retention and is expressed as grams of water required to saturate 100 grams of dry soil (William et al.). The high values of SP (76.6 and 111%) are an indication that the soil can store water. This confirms that there is no deep percolation in Gezira.

The inter particle ponds are also affected by the PH of the pore fluid in the sediment matrix, which affects the particle orientation during bonding by changing the surface or edge charges of the particles. A low PH ≤ 5.5 results in a stratified sediment bed exhibiting a high erosion rate near the surface. An intermediate PH (5 ≤ PH ≤ 7) results in a weaker bed structure due to lack in surface contact, which leads to more susceptibility to erosion. At high PH values when PH > 7, similar to our case, particle orientation predominates, the surface attraction forces become significant and form denser aggregates. This in agreement with Ravisangar *et al.* (2005) who observed that acidic water produces stronger sediment, which it does not erode easily due to excess shear stress.

The total amount of soluble soil ions can be estimated by measuring the electrical conductivity (EC) of a soil water extract. It is a significant factor in the susceptibility of cohesive soil to erosion (Sherard *et al.*, 1972). Soil with EC > 4 dS/m is saline soil. The results of the chemical analysis for the two samples in Table 5.21 show a low value of EC, which indicates that the sediment is not saline. The increase of the electrolyte concentration or change in the pore fluid of clay of a higher value tends to cause de-flocculation.

The ratio of dissolved sodium ions to the other main basic cations (calcium and magnesium) is known as the Sodium Adsorption Ratio (SAR). At high SAR, repulsive forces between particles are thought to be dominant, producing significant swelling and causing the clay to de-flocculate. Partheniades (2009) showed that decreasing values of SAR decreases the erosion rate. The analysis of the results show that the SAR has low values: 1.1 and 0.8. Aggregate stability (dispersion and flocculation) depends on the balance (SAR) between calcium (Ca^{2+}) and magnesium (Mg^{2+}), and sodium (Na^+) as well as the amount of soluble salts (EC) in the soil as depicted in Figure 5.50. The SAR can be obtained from Equation (5.3). Soil particles will flocculate if concentrations of $Ca^{2+} + Mg^{2+}$ are increased relative to the concentration of Na^+ (SAR is decreased). The soil is classified as normal soil according to Table 5.22.

$$SAR = \frac{Na^+}{\sqrt{(Ca^{+2} + Mg^{+2})/2}} \tag{5.3}$$

Table 5.22. Classification of soil based on electrical conductivity and SAR values

Soil classification	EC (dS/m)	SAR	Condition
Normal	< 4	<13	Flocculation
Saline	>4	<13	Flocculation
Sodic	<4	>13	Dispersed
Saline-Sodic	>4	>13	Flocculation

(Source: Walworth, 2006)

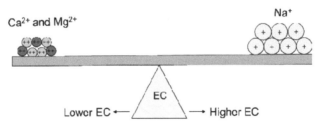

Figure 5.50. Distinction between flocculated and dispersed soil
(Walworth, 2006)

The exchangeable sodium percentage (ESP) is the presence of excessive amounts of exchangeable sodium that reverse the process of aggregation and cause soil aggregates to disperse into their constituent individual soil particles. This is known as deflocculation and occurs in sodic soil. The results show a very low percentage ESP < 6% and that the soil is non sodic. The ESP can be computed based on the following relation:

$$ESP = Exchangeable\ (Na\ /(Ca+Mg+K+Na))*100 \qquad (5.4)$$

It has been concluded that the type of sediment in Gezira Scheme has the properties of cohesive sediment and is fairly resistant to erosion.

5.6.5 Water column test

A series of laboratory tests were carried out to estimate the time that is needed for the sediment to settle. In these tests different water columns with different cylinder heights were used. This test was repeated for different sediment concentrations. The aim of these tests was to understand the mechanism of sediment in Gezira Scheme and to investigate the relationship between the concentration and time for the sediment to be deposited.

The amount of the sediment that will be deposited at different times (6, 12, 16, 18, 24 hours) has been detected. Consequently, it is possible to know how much sediment will be deposited if the water is stored in the minor canals for a certain period to determine the optimum time for the operation with less sediment. A sample was taken from upstream Toman offtake. It was found that most of the deposition occurs in the first 12 hours and the total deposition can occur within 24 hours. The relation between time of deposition and water column height is explained in Figure 5.51. More sediment is expected to be deposited for high concentrations. It was found that more than 73% of the sample was deposited in 6 hours when the concentration was 8390 ppm. Taking into consideration that the experiments were done in stagnant water by assuming that the velocity in the minor canal is very low and that all field canal gates were closed. The worst case can be obtained when the minor canal is closed due to rainfall and all the field canals are closed because of excess water for 6 hours or more so most of the sediment in the canal will be deposited.

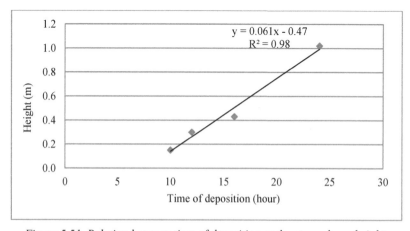

Figure 5.51. Relation between time of deposition and water column heights

Figure 5.52 displays cylinders with 35 cm height filled with water from Gezira Main Canal (kilo 77) with 2,100 ppm concentration. It is clear that the sediment is more homogenous in the water column and it seems that it has more wash load. No flocculation process is recognized, even not at high concentration.

t=0   t=6 hours   t=12 hours   t=16 hours   t=18 hours   t=24 hours

Figure 5.52. Process of deposition with time (hour)

5.7   Maintenance activities during the study period

In the past the maintenance activities were generally practiced between April and June during the canal closure period. Routine maintenance is used to be conducted in Gezira Scheme on a regular basis every two to three years depending upon the sedimentation in the canals. The longitudinal and cross-sectional surveys of the canals were conducted to quantify the amount of sediment, detecting the changes in canal prisms and bed slope referred to in the design. The absence of this systematic clearance due to financial limitation leads to more sediment accumulation in the canal prisms and rise in the bed level far above the design levels as explained in Sub-section 5.3.3.

The improper desilting campaigns or clearance activities have changed the geometry of the canals and the actual cross-sections far from the design profiles as depicted in Figure 5.53. The slope of Zananda Major Canal became 13 cm/km and 18 cm/km for the first and second reaches respectively. However, the measured bed profile in 1989 indicated that the slopes were 14.9 cm/km for the first reach and 7.6 cm/km for the second reach (HR Wallingford, 1990a), while the design slopes were 10 cm/km and 5 cm/km for the first and second reaches respectively. Steepening of the slope especially in the second reach is an indicator of improper desilting campaigns.

Large pools existed upstream of the control structures due to more excavation from the canals and banks. The berms of the canals have disappeared from the sections. There is an enlargement in Toman Minor Canal cross-section in comparison with the design section which has significantly contributed to a surge of deposition.

Due to limited financial resources during the last years, the priority of the clearance activities is given to the worse conditions and restricted to the emergence situation. The routine maintenance changed to emergency maintenance. In consequence,

water is delivered inadequately and inequitably to the water users. The following
maintenace activities during the study period were detected. In September 2011 the
group regulatours at K 9.1 and 12.5 were excavated in addition to the scond and third
reaches of Znanda Major Canal. In Toman Minor Canal the first, third and fourth
reaches were excavated. In 2012 the water user association took the responsibility of
excavating the second reach. Furthermore, the Sudan Gezira Board arranged clearance
work along Zananda Major Canal at the end of September without bathymeric survey
before or after the clearance. Long reach excavators were used. Cross-section surveys
for some sections have been arranged to assess the desilting activities. The results show
that the changes in cross-sections are not significant as shown in Figure 5.54.

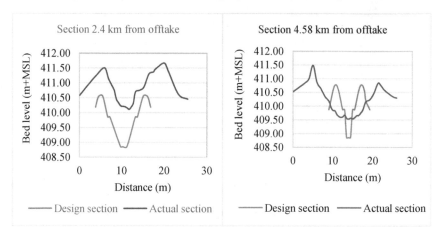

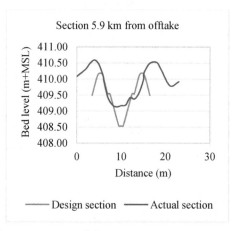

*Figure 5.53. Actual cross-section and design cross-section at some selected locations in
Toman Minor Canal*

5.8   Concluding remarks

Based on the obtained results the following concluding remarks can be made:
- instability of water levels along the canals was detected as a result of changes in
  the gate settings in a non-systematic mode;

- the bathymetric survey shows an increase in the bed level far above the design level. As a result, the water level became above the FSL (working level);
- more water is delivered to the system during most of the time in comparison with the crop water requirement due to the absence of systematic operation rules;
- the highest rate of sedimentation occurs over a short period between July and August. This coincides with the period when the largest sediment concentrations are transported by the Blue Nile River;
- the physico-chemical analysis indicated that the sediment has the properties of cohesive sediment;
- improper clearance activities have enlarged the canal cross-sections and reduced the water depth of the minor canals;
- the variety of crops within a Nimra and the absence of irrigation schedules have led to improper water management within the areas supplied by the minor canals;
- the collected data were used to calibrate and validate the developed model.

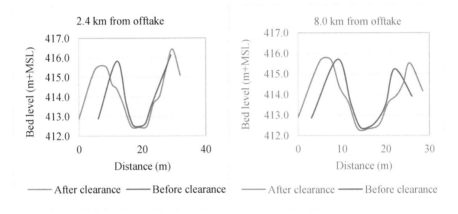

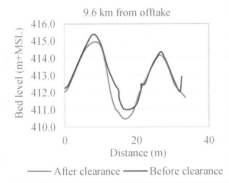

Figure 5.54. Cross-sections before and after the desilting at 400, 800 and 9600 m from the offtake on September 2012

6 Numerical model development

Most of the models that simulate sediment transport processes are mainly intended for rivers, estuaries and coastal areas. They are generally not developed with the specific characteristics of irrigation schemes. There is also a limitation in the models that can simulate cohesive sediment transport in irrigation canals. Although there are some similarities between rivers and irrigation canals, irrigation canals are different. The presence of a large number of flow control structures and the high influence of the side banks on the velocity distribution create some differences in both types of channels. The properties of the cohesive sediment and flow conditions (hydrodynamic process) imply that the prediction of the fine sediment is very difficult. The use of numerical models increases the understanding of the dynamics of fine sediment. Hence, it is important to develop a model dealing with fine sediment in irrigation canals that incorporates different types of hydraulic structures. The model for this study was developed by using Matlab software. Matlab is a high level language for numerical computations, programming and visualization, which has been used to simulate the sediment movement in irrigation canals. The model has been created by using the facilities provided by Matlab software such as interpolation, regression, integration and ordinary differential equations. The mathematical functions in Matlab support engineering and science operations, allow matrix manipulations and plotting of functions and data. The model is provided with movable weirs (adjustable crest levels) in addition to the other types of hydraulic structures that cannot be found in other models. A description of the concept of the model is presented in Figure 6.1.

6.1 Hydrodynamic computation

The inflow is given to the model in time series. The flow in irrigation canals was assumed steady non-uniform flow during the time step. The flow is steady since the flow rates of the outlets do not change with time and is non-uniform since the depth changes with location over the entire reach due to the presence and operation of the control structures and outlets. The canals were divided into reaches. The model computes the discharge at each point by applying the continuity equation. Then the water surface profile is predicted by solving the gradual varied flow equation numerically by using the predictor corrector method. The computation starts reach by reach from downstream towards upstream. The model computes the normal depth, critical depth, water depth and shear stress by assuming a logarithmic velocity distribution. The steps of computation are described in the next Sub-sections.

6.1.1 The flow computation

The reach is divided into Δx from x_1 to x_n. The discharge is determined starting from known discharge at the offtake of the major and minor canals (input data) based on the continuity equation. The discharge is withdrawn from the canals by laterals (Q_{lat}) such as minor canals or field canals in case of major or minor canals respectively. Figure 6.2 shows how the inflow Q_{in} from a supply canal is bifurcated over the offtake canals Q_{lat} and the continuing supply canal Q_{out} over distance Δx. The discharge after the lateral canals is following Equation (6.1). The outflow for each Δx is inflow for the next one.

$$Q_{out} = Q_{in} - Q_{lat} \qquad\qquad (6.1)$$

Where:

Q_{out} = outflow for each Δx (m³/s)

Q_{in} = inflow for each Δx (m³/s)

Q_{lat} = lateral discharge at Δx (m³/s)

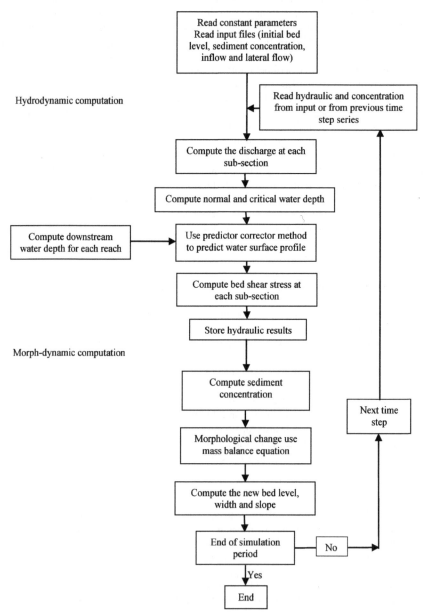

Figure 6.1. Computation processes of the model

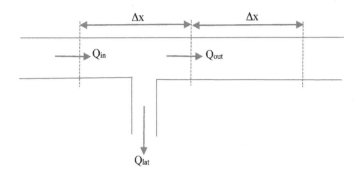

Figure 6.2 Flow balance sketch

6.1.2 Normal depth computation

The normal depth is computed by using Manning formula for each time step:

$$Q = \frac{AR^{2/3}S_f^{1/2}}{n} \tag{6.2}$$

Where:
Q   = discharge (m³/s)
A   = cross-section area of the canal (m²)
R   = hydraulic radius (m), R = A/p, p is the wetted parameter (m)
S_f = energy slope (-) equal to bed slope S_0 for uniform flow
n   = Manning coefficient (s/m$^{1/3}$)

For a trapezoidal section:

$$A = (b + my)y \tag{6.3}$$

Where:
A   = cross-section area (m²)
m   = side slope (-)
y   = water depth (m)
b   = bed width (m)

The wetted parameter for the trapezoidal section can be determined as:

$$p = b + 2y\sqrt{(1+m^2)} \tag{6.4}$$

Substitute all the above in Equation (6.2) gives:

$$Q = \frac{((b+my_n)y_n)^{5/3}S_f^{0.5}}{n(b+2y_n\sqrt{1+m^2})^{2/3}} \tag{6.5}$$

The normal depth (y_n) at each x is obtained from the above equation by using the Newton-Raphson method:

$$f(y_n) = ((b + my_n)y_n)^{5/3} S_0^{0.5} - Qn(b + 2y_n \sqrt{1 + m^2})^{2/3} \tag{6.6}$$

The first derivative is as follows:

$$f'(y_n) = \frac{5}{3}((b+my_n)y_n)^{2/3} S_0^{0.5} *(b+2my_n) - \frac{2}{3}nQ(b+2y_n\sqrt{1+m^2})^{-1/3}(2\sqrt{1+m^2}) \tag{6.7}$$

Newton-Raphson method is applied to get y_n:

$$y_{n_2} = y_{n_1} - \frac{f(y_{n_1})}{f'(y_{n_1})} \tag{6.8}$$

6.1.3 Critical depth computation

The Froude number (Fr) is defined as:

$$Fr = \frac{Q}{\sqrt{g\dfrac{A^3}{T}}} \tag{6.9}$$

Where:
Fr = Froude number (-)
Q = discharge (m³/s)
g = gravity acceleration (m/s²)
T = top surface width (m)
A = cross-section area (m²)

Critical flow occurs when Fr = 1, then:

$$Q^2 = \frac{gA^3}{T} \tag{6.10}$$

The above equation can be written as:

$$Q^2 = \frac{(b+my_c)^3 y_c^3 g}{(b+2my_c)} \tag{6.11}$$

Equation (6.11) is solved to get y_c by using the Newton-Raphson method:

$$f(y_c) = (b + my_c)^3 y_c^3 g - Q^2 (b + 2my_c) \qquad (6.12)$$

The first derivative is as follows:

$$f'(y_c) = 3g((b + my_c)y_c)^2 (b + 2my_c) - 2mQ^2 \qquad (6.13)$$

The Newton-Raphson method is applied to get y_c for each x_i:

$$y_{c_2} = y_{c_1} - \frac{f(y_{c_1})}{f'(y_{c_1})} \qquad (6.14)$$

6.1.4 Water surface profile

The water flow calculations are based on sub-critical, quasi-steady flow. It is assumed that the changes in flow profiles due to operation of the gates or change in demand are gradual and that the duration of these changes is considerably short as compared to the duration of the irrigation season and that its effects on the bed change from seasonal perspective is insignificant. The intermediate structures of the major canals are weirs thus the flow depth at the weirs is critical. The computation starts at the end of the reach and proceeds upstream as mentioned before. Based on the water depth, critical depth and normal depth, the water surface profiles are classified. It was found that the water surface profile is Mild (either M1 or M2).

Starting from the total energy equation (Chow, 1993)

$$H = z + y + \frac{u^2}{2g} \qquad (6.15)$$

The differentiation with respect to distance is computed as follows:

$$\frac{dH}{dx} = \frac{dz}{dx} + \frac{dy}{dx} + \frac{d(u^2/2g)}{dx} \qquad (6.16)$$

Where:

$$\frac{dH}{dx} = -S_f \qquad (6.17)$$

$$\frac{dz}{dx} = -S_0 \qquad (6.18)$$

Based on a known flow rate:

$$\frac{d(u^2/2g)}{dx} = -\frac{Q^2 T}{gA^3}\frac{dy}{dx}$$ (6.19)

Equation (6.16) can be written as follows:

$$(\frac{dy}{dx})_i = \frac{S_0 - S_f}{1 - Fr^2}$$ (6.20)

Where:
H   = total head (m)
u   = flow velocity (m/s)
z   = bed elevation (m)
dy/dx = water surface profile slope (-)
S_f = friction slope (-)
S_0 = bed slope (-)
Fr   = Froude number (-)
T   = top surface width (m)

There are several methods to solve the gradually varied flow equation for prismatic canals. Henderson (1996), Chow (1993) and Depeweg (1993) review clearly the available methods such as:
- direct integration by using the Bresse method or the Backhmeteff method;
- graphical integration method;
- numerical integration (predictor-corrector method, standard step method and direct step method).

The predictor-corrector method is used to calculate the flow profile. The accuracy of the prediction of the water depth in the model is 0.005 m. The model starts from known parameters downstream such as: water depth (y_0) at the downstream boundary, discharge (Q), roughness coefficient (Manning coefficient (n)), bed level (z_0), bed slope (S_0), bed width (b) and side slope (m). Figure 6.3 shows the starting point of the computation. The computation of the water surface profile by applying the predictor-corrector method is shown in Figure 6.4.

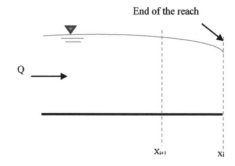

Figure 6.3. Starting point of the water surface profile computation

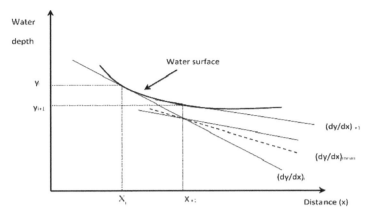

*Figure 6.4. Predictor-corrector method for water surface profile computation
(Munir, 2011)*

The steps of the computation are:

i.   from known y_i the energy slope can be computed by using Equation (6.2);
ii.   the Froude number is computed by using Equation (6.9);
iii.   by using the momentum equation for varied flow at the first point i
       (downstream) and by obtaining dy/dx from Equation (6.20);
iv.   for the next step, y_{i+1} is computed as follows:

$$y_{i+1} = y_i + (\frac{dy}{dx})_i (X_{i+1} - X_i)$$

(6.21)

v.   the above steps are repeated to get A_{i+1}, R_{i+1}, S_{fi+1}, Fr_{i+1}
vi.   from Equation (6.20) $(\frac{dy}{dx})_{i+1}$ is computed and then $(\frac{dy}{dx})_{mean}$ from the
       following equation:

$$(\frac{dy}{dx})_{meam} = \frac{(\frac{dy}{dx})_{i+1} + (\frac{dy}{dx})_i}{2}$$

(6.22)

vii.   the new water depth is calculated by the following equation:

$$\overline{y}_{i+1} = y_i + (\frac{dy}{dx})_{mean} (X_{i+1} - X_i)$$

(6.23)

viii. the accuracy of the predictor-corrector method is checked by:

$$\left| y_{(i+1)} - \overline{y}_{(i+1)} \right| \le e$$

(6.24)

Where:
e = degree of accuracy: 0.005 m

The above steps have to be repeated till $X_i = X_n$ at the head of the reach.

6.1.5 Shear stress computation

A bed formed by deposition of fine sediments, which is normally the case in estuaries and other conduits with cohesive boundaries in which fine sediment is transported and deposition takes place, can be reasonably assumed as smooth (Partheniades, 2009). Once the canal has uniform flow, the turbulent boundary layer will be fully developed. Therefore, the velocity distribution will have a definite pattern (Chow, 1993). It is expected that in a cohesive sediment environment, where the bed structure has cohesive properties, the bed is often hydraulically smooth (Partheniades, 2009). Therefore, it was assumed that the roughness is small and does not affect the velocity distribution outside the laminar sub-layer and the flow is hydraulically smooth. Simons and Senturk (1992) emphasized that in pipes and open channels, the velocity distribution is given by the well-established logarithmic law for smooth boundaries:

$$\frac{u}{u_*} = 5.75\log\frac{u_* y}{v} + 5.5 \tag{6.25}$$

Where:

u_* = shear velocity (m/s)
u = flow velocity at y depth measured from the bed (m/s)
v = kinematic viscosity (m²/s)

For a trapezoidal section:

$$dQ = uTdy \tag{6.26}$$

$$dQ = u(b + 2my)dy \tag{6.27}$$

Where:
Q = discharge (m³/s)
T = top surface width (m)
b = bed width (m)
y = depth from the bed (m)
m = side slope (-)

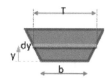

The integration implies the substitution of Equation (6.25). The integration starts from 0.025y above the bed level of the canal where the velocity is very low near the bed. The reduction of this value does not significantly affect the results.

$$Q = b \int_{0.025y}^{y} udy + 2m \int_{0.025y}^{y} uydy \tag{6.28}$$

$$Q = u_* b \int_{0.025y}^{y} 5.75\log\frac{u_* y}{\gamma}dy + 5.5u_* b \int_{0.025y}^{y} dy + 2mu_* \int_{0.025y}^{y} (5.75\log\frac{u_* y}{\gamma})ydy + 11mu_* \int_{0.025y}^{y} ydy \tag{6.29}$$

By using the Newton-Raphson method the equation is solved for u_* :

$$f(u_*) = u_* b \int_{0.025y}^{y} 5.75 \log \frac{u_* y}{\gamma} dy + 5.5 u_* b \int_{0.025y}^{y} dy + 2mu_* \int_{0.025y}^{y} (5.75 \log \frac{u_* y}{\gamma}) y dy + 11 mu_* \int_{0.025y}^{y} y dy - Q$$

(6.30)

$$u_{*2} = u_{*1} - \frac{f(u_{*1})}{f'(u_{*1})}$$

(6.31)

Where f' is 1st derivative of f, u_{*1} is the starting value and u_{*2} is the computed value, solved till $(f(u_{*1}) / f'(u_{*1})) < 0.005$. Some of the algebraic computations such as integration and derivation are being done by mathematical software.

Then the shear stress is computed at each x_i as follows:

$$\tau = \rho_w u_*^2$$

(6.32)

Where:

τ = shear stress (N/m^2)

ρ_w = water density (kg/m^3)

6.2   Suspended sediment transport computation

The model uses an uncoupled solution. First, the water flow equations are solved (by using the predictor-corrector numerical method). Then, the results are used as an input to solve the sediment transport equation. It was assumed that all the sediment is transported as suspended and the transported bed load can be neglected since most of the sediments are fine sediments. The estimation of suspended sediment transport in streams is mainly based on the theory of the energy approach and the diffusion-dispersion theory (including exchange theory). However, the diffusion-dispersion theory is recommended over the energy theory since experimental evidence indicates that it better fits with observed data (Simons and Senturk, 1992). The suspended sediment concentration is calculated based on solving the one dimensional advection diffusion equation (Huang et al., 2008):

$$\frac{\partial C}{\partial t} + u \frac{\partial C}{\partial x} = \frac{1}{A} \frac{\partial}{\partial x} (A K_s \frac{\partial C}{\partial x}) + S$$

(6.33)

Where:

u   = average flow velocity (m/s)

C   = average sediment concentration (kg/m^3)

A   = cross-section area (m^2)

K_s = longitudinal dispersion coefficient (m^2/s)

S   = source/ sink term (kg/m^2/s) per unit length of channel

The source/sink term can be written as follows:

$$S = M(\frac{\tau}{\tau_c} - 1) - w_s C \qquad (6.34)$$

The concentration C is obtained by solving the one dimension advection-diffusion equation numerically. The above equation is re-written as follows:

$$\frac{\partial}{\partial t}(CA) + \frac{\partial}{\partial x}(CQ) = \frac{\partial}{\partial x}(A K_s \frac{\partial C}{\partial x}) + SA \qquad (6.35)$$

The finite difference method is used to solve the differential equation. The domain is partitioned in space $(x_1, x_2...,x_n)$ and in time t_k (k = 0,1,2,...) by using a mesh and assuming uniform partitions in space and time (Δx and Δt) as shown in Figure 6.5. A forward difference at time t_k and a second-order central difference for the space derivative at position x_i is adopted with exception of the first and last points (i.e. x_1 and x_n) where a second-order forward difference and a backward difference is used.

The forward difference for the temporal derivative is as follows:

$$\frac{\partial f(t_k)}{\partial t} = \frac{f(t_k) - f(t_{k-1})}{\Delta t} \qquad (6.36)$$

The spatial derivative in the central method for equal space discretization is given by:

$$\frac{\partial f(x_i)}{\partial x} = \frac{f(x_{i+1}) - f(x_{i-1})}{2\Delta x} \qquad (6.37)$$

$$\frac{\partial^2 f(x_i)}{\partial x^2} = \frac{f(x_{i-1}) - 2f(x_i) + f(x_{i+1})}{\Delta x^2} \qquad (6.38)$$

The forward difference is given by:

$$\frac{\partial f(x_i)}{\partial x} = \frac{-3f(x_i) + 4f(x_{i+1}) - f(x_{i+2})}{2\Delta x} \qquad (6.39)$$

$$\frac{\partial^2 f(x_i)}{\partial x^2} = \frac{2f(x_i) - 5f(x_{i+1}) + 4f(x_{i+2}) - f(x_{i+3})}{\Delta x^2} \qquad (6.40)$$

The backward difference is given by:

$$\frac{\partial f(x_i)}{\partial x} = \frac{f(x_{i-2}) - 4f(x_{i-1}) + 3f(x_i)}{2\Delta x} \qquad (6.41)$$

$$\frac{\partial^2 f(x_i)}{\partial x^2} = \frac{-f(x_{i-3}) + 4f(x_{i-2}) - 5f(x_{i-1}) + 2f(x_i)}{\Delta x^2} \qquad (6.42)$$

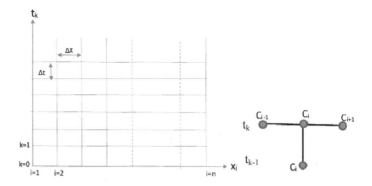

Figure 6.5 Computational grid of the finite difference method

At a certain time t_k, Equation (6.33) can be discretized and written in a matrix-form as:

$$[B]\{C\}=\{D\}$$ (6.43)

[B] is a band matrix and its nonzero elements for t_k are given by:

$$B_i = -\beta_i - \frac{A_i}{\Delta t} - 2\frac{A_i K_s}{(\Delta x)^2}$$ (6.44)

$$B_{i+1} = \frac{A_i K_s}{(\Delta x)^2} + \frac{K_s}{2\Delta x}\left(\frac{\partial A}{\partial x}\right)_i - \frac{Q_{i+1}}{2\Delta x}$$ (6.45)

$$B_{i-1} = \frac{A_i K_s}{(\Delta x)^2} - \frac{K_s}{2\Delta x}\left(\frac{\partial A}{\partial x}\right)_i + \frac{Q_{i-1}}{2\Delta x}$$ (6.46)

$$D_i = -\frac{(C_i)_{k-1} A_i}{\Delta t} - \chi_i$$ (6.47)

Where:

$$\beta_i = \frac{A_i w_{si}}{\Delta x}$$ (6.48)

$$\chi_i = \begin{cases} \dfrac{A_i M}{\Delta x}\left(\dfrac{\tau_i}{\tau_c} - 1\right) & \text{if } \tau_c < \tau \\ 0 & \text{if } \tau_c \geq \tau \end{cases}$$ (6.49)

The spatial domain is discretized into n points and i in the above equations that takes the values between 2 and n-1; $2 \leq i \leq n-1$.

The nonzero elements in the last row of matrix B are given by:

$$B_n = -\beta_n - \frac{A_n}{\Delta t} + \frac{2A_n K_s}{(\Delta x)^2} + \frac{3K_s}{2\Delta x}\left(\frac{\partial A}{\partial x}\right)_n - \frac{3Q_n}{2\Delta x} \tag{6.50}$$

$$B_{n-1} = -\frac{5A_n K_s}{(\Delta x)^2} - \frac{2K_s}{\Delta x}\left(\frac{\partial A}{\partial x}\right)_n + \frac{2Q_{n-1}}{\Delta x} \tag{6.51}$$

$$B_{n-2} = \frac{A_n K_s}{(\Delta x)^2} + \frac{K_s}{2\Delta x}\left(\frac{\partial A}{\partial x}\right)_n - \frac{Q_{n-2}}{2\Delta x} \tag{6.52}$$

$$B_{n-3} = -\frac{A_n K_s}{(\Delta x)^2} \tag{6.53}$$

The nonzero elements in the matrix can be written as $B_{i,j}$, where i refers to the row and j to the column.

At time k, the concentration for the first point $(C_1)_k$ is obtained by interpolating the field measurements. For the other points it is given by:

$$
\begin{Bmatrix} C_2 \\ C_3 \\ \bullet \\ \bullet \\ \bullet \\ C_n \end{Bmatrix}_t
=
\begin{bmatrix}
B_{2,2} & B_{2,3} & & & & \\
B_{3,2} & B_{3,3} & B_{3,4} & & & \\
& \bullet & \bullet & \bullet & & \\
& & \bullet & \bullet & \bullet & \\
& & & \bullet & \bullet & \bullet \\
& & B_{n,n-3} & B_{n,n-2} & B_{n,n-1} & B_{n,n}
\end{bmatrix}^{-1}
\begin{Bmatrix} D_2 - (C_1)_t B_{2,1} \\ D_3 \\ \bullet \\ \bullet \\ \bullet \\ D_n \end{Bmatrix}
\tag{6.54}
$$

The model has been developed for one sediment fraction; hence the factor k is incorporated in the model and given a unit value. It can be given different values according to different fractions. The effect of hinder in the settling velocity is considered in the model when the concentration is above 10,000 ppm (Partheniades, 2009). The settling velocity varies spatially and temporally according to the concentration and is obtained by Equation (5.1). The effective settling velocity is obtained by Equation (4.6). Reynolds number of the particle (Re_p) was computed from Equation (4.7) and it was found that it has a value of 0.073 for an average settling velocity (w_s) of 0.61 mm/s for the 90% sediment fraction, mean particle size (d_{50}) of 0.12 mm and kinematic viscosity (v) taken at 10^{-6} m^2/s. Then, the exponential (n) was taken as 4.7 from Figure 4.2.

To compute the sediment concentration at t_0 when k = 0, the one dimensional convection-diffusion equation has been simplified; the diffusion term is ignored since the dimensional analysis showed that the diffusion term can be neglected in steady state solutions (Munir, 2011). Uniform supply of sediment was assumed ($\frac{\partial C}{\partial t} = 0$).

Accordingly, the simple form of Equation (6.35) represents the mass balance approach that is applied for the computation of the deposition and erosion/re-suspension.

$$\frac{\partial(CQ\ )}{\partial x} = S \times A \tag{6.55}$$

The above equation can be written in mass balance form:

$$Q_{s_{i+1}} = Q_{s_i} - D_i A_i + E_i A_i \tag{6.56}$$

$$Q_{si} = Q_i C_i \tag{6.57}$$

Where:

Q_{s_i} = upstream sediment load (kg/s)

$Q_{s_{i+1}}$ = downstream sediment load (kg/s)

S   = exchange rate between bed and water in (kg/m^2/s) per unit length
D_i = deposition flux (kg/m^2/s)
E_i = erosion flux (kg/m^2/s)
A_i = cross-section area (m^2)
Q_i = discharge (m^3/s)

At i = 1 and k = 0 the inflow and sediment concentration are given as inputs. Therefore, Equation (6.55) is solved for C_{i+1} since the discharge is known at each Δx from the continuity equation. The source/sink term represents the water-bed exchange rate and takes into account the erosion and deposition of sediment per unit length. The cohesive sediment model uses the bottom shear stress as a governing factor for the erosion as well as deposition processes. The deposition is calculated as the flux of sediment towards the bed as formulated by Krone (1962) and the erosion/re-suspension from the bed are based on Partheniades (1965), Equation (4.2) and Equation (4.13) respectively. The re-suspension occurs when the bed shear stress is too large. Particles in the sediment bed can be reworked because of the physical and biological mixing processes. The physical mixing process near the bed surface can be considered as a diffusion process with a certain mixing coefficient and increases with the increase of the velocity (Ledden, 2003).

There are two conditions for erosion and deposition; if the bed shear stress is less than the critical shear stress for erosion, no erosion will occur ($\tau_c \geq \tau$, $E = 0$). If the bed shear stress is greater than the critical shear stress for deposition; no deposition will occur ($\tau_d \leq \tau$, $D = 0$). The behaviour of critical shear stress for a full deposition or a partial deposition is not well understood but the accuracy of the deposition in the model mainly depends on the use of the correct values. Therefore, it can be a calibration parameter for determining the deposition rate (Huang et al., 2008). With the poor equipped labs in Sudan, it was very difficult to check the existence of the critical shear stress of deposition. However, the critical shear stress for deposition depends on the flow condition as well as sediment properties and it differs from one place to another. The approach of the critical shear stress of deposition is followed as mentioned in Sub-section 4.4.4.

For uniform flow, the diffusion coefficient was assumed to have a constant value. At low concentration, the variation in flocculation over the water depth or with time does not play an important role (Krone, 1962, Winterwerp and Kesteren, 2004). However, at high concentration, it has an essential effect in the floc size and hence on the particle density over the water depth. Based on this, the effect of the flocculation in the settling velocity is not taken into account. The settling velocity is considered to be constant over the water depth since 92% of the measured concentration is less than 10,000 ppm and only exceeds this value in 4 to 5 days during the flood period.

The average concentration has been selected instead of the concentration near the bed in Equation (4.2). It was assumed that the sediment enters the canal per unit time and is immediately spread uniformly over the entire section. Figure 5.35 shows that the fine sediment distribution is nearly uniform over the water column when the vertical distribution of the sediment in Gezira Scheme is plotted in the graphical representation of the Rouse Equation. In case of outflow (lateral canals) in the system, the concentration at the offtake remains the same as in the parent canal, since the variation in the values is not significant according to the measured data.

6.3   Morphological changes

For a small Froude number (Fr < 0.4), the celerity of the water movement is much larger than the celerity of the sediment movement in the bed (De Vries, 1975). Though the changes in the flow from one stage to another are so fast that the bed change during that period can be assumed to be constant (De Vries, 1975). In other words, the time scale of bed change is much smaller than that of flow movement. Therefore, at each time step the flow is calculated assuming a "fixed" bed. The change in the bed level is computed by solving the mass balance equation or sediment continuity (Equation 6.58) explicitly and numerically by using the finite difference method. For upstream and downstream boundaries the forward and backward method was used and in between the central difference method was used (Munir, 2012).

$$\frac{\partial z}{\partial t} + \frac{1}{b\rho}\frac{\partial Q_s}{\partial x} = 0 \qquad\qquad (6.58)$$

Where:

Q_s = sediment load (kg/s)

b   = bed width (m)

ρ = sediment density (kg/m^3)

z   = bottom level above datum (m)

x   = length coordinates in x direction (m)

ρ = sediment density (kg/m^3)

Upstream boundary:

$$z_{i,j+1} = z_{i,j} - \frac{\Delta t}{b\rho}\frac{Q_{s_{i+1,j}} - Q_{s_{i,j}}}{\Delta x} \qquad\qquad (6.59)$$

Downstream boundary:

$$z_{i,j+1} = z_{i,j} - \frac{\Delta t}{b\rho}\frac{Q_{s_{i,j}} - Q_{s_{i-1,j}}}{\Delta x} \qquad\qquad (6.60)$$

$$z_{i,j+1} = z_{i,j} + \frac{\Delta t}{b\rho} \frac{Q_{S_{i+1,j}} - Q_{S_{i-1,j}}}{2\Delta x} \qquad (6.61)$$

The stability of the solution is checked by using the Courant number. It should be noted that the Courant number should be small in order to reduce the oscillation, improve the accuracy and decrease the numerical dispersion (Munir, 2011).

$$C_r = \frac{u\Delta t}{\Delta x} \qquad (6.62)$$

Where:
C_r = Courant number (-)
u   = average linear velocity for each location (m/s)
Δx = dimension of grid cell at each location (m)
Δt = time step (s)

The time step in the model has been set at 100 seconds and the reach is divided into 201 nodes, therefore, Δx = 46 m to improve the accuracy of the results. The Courant number for maximum velocity of 0.54 m/s is about 1.17. Implicit solvers are usually less sensitive to numerical instability and larger values of the Courant number can be tolerated while for explicit solutions the Courant number should be one or less than one (Guo et al., 2013).

6.4   Boundary conditions

The upstream boundary conditions are the head discharge $Q(x_1, t_k)$ and the inflow sediment concentration $C(x_1, t_k)$. The downstream boundary conditions are the water levels $y(x_n, t_k)$ for each canal reach. The computation of the water level at the end of each reach is based on the characteristics of the hydraulic structures and the discharge passes through the control structures. The water depth at the tail of the canal is assumed to be at normal depth. It is also assumed that the water depth does not change during the time step.

6.5   Types of hydraulic structures included in the model

Different types of hydraulic structures in the model will be highlighted in this section. The locations of the structures define the boundaries between reaches.

Overflow control structures

The broad crested weirs in addition to other types of weirs are considered in the model based on Equation (3.7). The intermediate structures in most of the major canals are adjustable overflow structures (movable weir (MW-II)) as explained in Sub-section 3.1.5. The crest levels were introduced in the model in time series. The movable weirs in Gezira Scheme are designed for a modular limit of 0.7 as mentioned before. Figure 6.6 shows a typical overflow structure. The upstream water level is computed by:

$$Q = C_d w h_1^n \qquad (6.63)$$

$$z_u = z_w + (\frac{Q}{C_d w})^{1/n}$$

(6.64)

Where:

Q = discharge (m³/s)
w = width of the weir (m)
z_u = upstream water level above mean sea level (m+MSL)
z_w = crest level above sea mean level (m+MSL)
n = exponent (-)
C_d = discharge coefficient (-)
h_1 = upstream water depth above the crest (m)
h_2 = downstream water depth above the crest (m)

In the major canal under study this formula was used with $C_d = 2.3$, n = 1.6 for cross structure-1 and $C_d = 2.17$, n = 1.35 for cross structure-2 as explained in Annex D. The case of overflow above the top of the fixed beam is also considered in the model by treating the fixed beam as a sharp crested weir (Annex D).

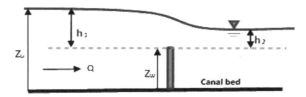

Figure 6.6. Typical overflow structure

Drop structure

Drop structures are mostly built in watercourses with steep slopes and are designed to pass the water to a lower elevation while controlling the energy and velocity of the water as it passes over. This type of structure is incorporated into the model when the canal bed level upstream of the overflow structure becomes above the crest level as a result of sedimentation. Thus, the movable weir becomes out of function. In this study, the bed level at the second reach upstream of the control structure was at 410.40 m+MSL while the crest level at its lowest position was 410.35 m+MSL. The control structure is treated as a drop structure for most of the time when the bed level upstream was above the crest level. The downstream water depth is computed by assuming critical flow and a rectangular section at the edge of the drop structure as shown in Figure 6.7. Equation (6.65) is applied for critical flow:

$$y_c = \sqrt[3]{\frac{q^2}{g}}$$

(6.65)

Where:

y_c = critical water depth (m)
q = discharge per unit width (m²/s)
g = gravity acceleration (m/s²)
Fr = Froude number (-)

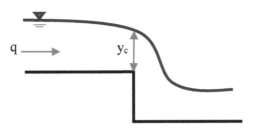

Figure 6.7. Drop structure

Undershot structure

Pipe regulators can be found in the minor canals in Gezira Scheme as cross structures. The flow through a pipe regulator is treated as orifice since the length of the pipe is short and the friction losses are not considered. Figure 6.8 presents the two flow cases for an undershot structure. The discharge is more dependent on the head difference between upstream and downstream for submerge flow. The relation for submerged flow is as follows (Rajaratnam and Subramanya, 1967).

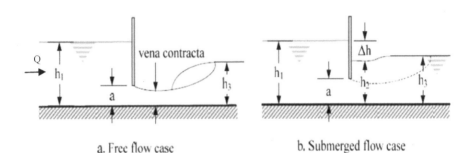

a. Free flow case   b. Submerged flow case

Figure 6.8. Undershot structure (Paudel, 2010)

$$Q = C_d A \sqrt{2g(h_1 - h_3)} \tag{6.66}$$

Where:

$$C_d = \frac{C_c}{\sqrt{1 - (\frac{C_c a}{h_1})^2}} \tag{6.67}$$

For free flow the downstream water depth should be:

$$h_3 < \frac{C_c a}{2} \left[\sqrt{1 + Fr^2} - 1 \right] \tag{6.68}$$

Then, the following relation should be applied:

$$Q = C_d A \sqrt{2gh_1 - C_c a}$$ (6.69)

Where:
Q   = discharge (m^3/s)
A   = opening area (m^2)
C_c = contraction coefficient that can be approximately 0.64
Fr   = Froude number corresponding to the downstream flow depth (-)
h_3 = downstream water depth (m)
h_1 = upstream water depth (m)
a = height of the orifice (m)

The discharge depends upon the upstream water depth and gate opening in case of free flow. The discharge coefficient is a function of (h_1/a) with a range between 0.596 and 0.607 (Bos, 1989). The contraction coefficient C_c is a function of (h_1/a) as well and has a value between 0.624 and 0.648. For fully contracted, submerged, rectangular orifices, the discharge coefficient C_d = 0.61 (Bos, 1989).

At the minor canal the model starts the computation from the tail of the canal by assuming that the water depth is at normal depth and the water depths is computed towards upstream by using the predictor corrector method. Equations (6.66) and (6.69) are solved for h_1 according to the flow condition, whether it is submerge or free flow.

7 Simulation of suspended sediment transport in irrigation canals

The developed model was setup, calibrated and validated based on the collected data. The effects of different parameters on the deposition were investigated to detect the most effective ones. The main objective of the modelling and simulation of suspended sediment transport in Gezira scheme was to analyse and investigate the responses of the system to sediment deposition under different operation and management scenarios.

7.1 Model setup, calibration, validation and verification

7.1.1 Model setup

The physical process of advective transport and diffusion determines the movement and change in the sediment concentration. There is a large variation in the longitudinal dispersion coefficient values for irrigation canals and rivers (Sahay, 2013). In fact there is no clear identified range of this coefficient in irrigation canals. Several studies have been carried out to present simple formulae for its prediction (Lui, 1977; Fischer *et al.*, 1979; Seo and Cheong, 1998; Deng *et al.*, 2001; Kashefipour and Falconer, 2002; Sahay, 2013). Since the formulae were developed for different conditions, the predicted values differ significantly. They all present the coefficient based on flow velocity, shear velocity, water depth and width of the canal. However, the predicted values are varying since the phenomenon of the dispersion is complex and it still remains a challenge to quantify this coefficient (Sahay, 2013). The sensitivity of the longitudinal dispersion coefficient to the deposition was tested. The model shows that the change in dispersion coefficient has no effect on the deposition but contributes significantly to the stability of the numerical solution. Stability was achieved when the dispersion coefficient was less than 10 m^2/s. Therefore, the longitudinal dispersion was taken as 10 m^2/s in this study.

The bed profile of the canals was loaded in a text file every 200 m based on the bathymetric survey. The abscissa of each reach, the minor canal locations as described in Table 5.3, the width and the side slope of each reach were defined in the model. The bed slopes were set 0.00013 and 0.00018 for the first and second reach respectively. The average bed width is 5 m for the first reach and 4 m for the second reach. The average side slope was (1:1.1) for the first reach and (1:1) for second reach. The discharge and sediment concentration were set in time series at the upstream boundary. The outflow along the canal during the simulation period was also set in time series. It was assumed that during the closure (rainfall period) the bed level remains constant and this period was excluded from the simulated period. The types of structures and their properties such as the discharge coefficient, width of weirs and exponential values (Section 6.5) were defined in the model. The sediment and flow properties were defined in the model such as particle density 2740 kg/m^3, water density 1000 kg/m^3 and kinematic viscosity 10^{-6} m^2/s. The dry sediment density of the silt deposits was taken as 1200 kg/m^3 (Ali, 2014) when the sediment transport of the Blue Nile River was studied. The approach of the critical shear stress of deposition is followed. Thus the critical shear stress for deposition is given the high value of 1000 N/m^2, which means that the suspended sediment is always allowed to deposit. The critical shear stress of erosion controlled the deposition and can be a calibrated parameter as well as the re-adaptation parameter (M).

7.1.2 Model calibration

Two types of calibration approaches have been considered in sediment transport modelling; the integrated and instantaneous approaches. The instantaneous approach can be obtained by evaluating the concentration formulae by using the actual concentration at different abscissa at the same time. In the integrated calibration the parameters in the formulae are related to the geometry of the canals. It consists in minimizing the difference between the simulated bed profiles and the measured one. According to Munir (2011) the integrated method is more reliable than the instantaneous method and more important for irrigation and maintenance management. Thus the integrated method was followed in this study.

The dynamics of cohesive sediment are described by theoretical empirical equations, which render modelling of the sediment difficult. Moreover, it is difficult to measure the important calibration parameters such as critical shear stress for erosion and erosion rate that makes the situation more complicated. In practice the calibration is to search for a reasonable combination of parameters that can make the measured bed profile fitting with the simulated profile. This does not grantee that other combinations of parameter settings can provide an equally good calibration. Therefore, it is important to understand the behaviour of the sediment dynamics to assess the reliability and reality of the predicted results. Munir (2010) proposed criteria to check the accuracy of the predicted values:

$$RMSE = \left[\frac{1}{n} \sum_{i=1}^{n} (M_i - S_i)^2 \right]^{0.5} \times \left(\frac{100}{M} \right) \tag{7.1}$$

$$ME = \max \left| M_i - S_i \right|_{i=1}^{n} \tag{7.2}$$

$$EF = \frac{\sum_{i=1}^{n} (M_i - M)^2 - \sum_{i=1}^{n} (M_i - S_i)^2}{\sum_{i=1}^{n} (M_i - M)^2} \tag{7.3}$$

$$MAE = \frac{1}{n} \sum_{i=1}^{n} \left| M_i - S_i \right| \tag{7.4}$$

Where M_i, S_i and M are the measured, simulated and average of measured values respectively.

The root mean square error (RMSE) is the total difference between simulated and measured values, proportional against the mean observed values. The RMSE should be less than half the standard deviation of the measured values to ensure accurate simulation. The maximum error (ME) indicates the maximum error between the simulated and measured values. Furthermore, the modelling efficiency (EF) is a measure for assessing the accuracy of simulations. The maximum value for EF is one when the measured values match the simulated ones perfectly. The mean absolute error (MAE) is the mean absolute error estimation. The best values of ME and MAE are close to zero.

The hydrodynamic part of the model was calibrated by adjusting the input parameters such as the roughness coefficient. According to Chow (1993) the resistance to flow in open channels depends on the surface roughness, vegetation, channel irregularity, sediment size, shape of channel and stage discharge. The bed roughness consists of the grain roughness related to the bed grain materials and roughness due to bed form. For the fine sediment in irrigation the bed form component has not been noticed. The grain roughness refers to the shear force created by the sediment particles at the boundary. It is more difficult to measure the roughness of the canal in the field (Halcrow Group, 1998). The common practice in this case is to estimate the range or the value by applying the Manning formula at different locations based on the flow measurements and the geometry of the canal cross-section. However, it was found from the results that the average of the Manning roughness coefficient was not varying much as revealed in Table 7.1. Thus a constant value was assumed. The model shows that the slight change in the roughness coefficient does not affect the deposition significantly as explained in Sub-section 7.1.6. Chow (1993) analysed the Manning coefficient for excavated weedy earth canals and found that it is in the range 0.022-0.033 $s/m^{1/3}$. When the bed material is fine the value of n is low and relatively unaffected by changes in the flow stage (Chow, 1993). The adjusted value of the Manning coefficient was found to be 0.023 $s/m^{1/3}$, which represents the average measured values in Table 7.1 when considering the weed growth. Gradual variations of the cross-section have a rather insignificant effect on the roughness coefficient (Khan, 2004). Then, the predicted water levels downstream of the offtake were compared with the measured values as revealed in Figure 7.1. The water levels were changed instantaneous according to the abstraction downstream and change in the water level upstream.

Table 7.1. Manning coefficient from field measurement

Discharge (m^3/s)	Bed width (m)	Side slope (-)	Water depth (m)	Bed slope (-)	Manning coefficient $(s/m^{1/3})$
4.60	6.00	1.10	1.18	0.00013	0.020
5.01	6.00	1.10	1.20	0.00013	0.019
5.15	5.50	1.10	1.32	0.00013	0.020
2.56	5.00	1.10	0.90	0.00013	0.019
2.36	4.00	1.00	1.00	0.00018	0.023
2.22	4.00	1.00	0.90	0.00018	0.020
1.69	4.00	1.00	0.85	0.00018	0.024
Average value					0.021

Winterwerp and Kesteren (2004) gave typical values $0.01\times10^{-3} < M < 0.5\times10^{-3}$ $kg/m^2/s$, while Ellegsequent and Christiansen (1994) gave M values between 0.00045 and 0.0016 $kg/m^2/s$. In this study, M was taken at 0.0016 $kg/m^2/s$ in the same order of magnitude as previously published results of Whitehouse et al. (2000) and Lumborg (2005). The sensitivity analysis shows that the change in the M parameter has an effect on the deposition. The critical bed shear stress for erosion and the erosion rate were generally assumed to be constant parameters although they vary with depth and time as the result of consolidation and physico-chemical effects. Generally the erosion rate (M) is taken as a constant (Winterwerp and Kesteren, 2004). The critical shear stress for erosion was the key parameter for estimating the amount of deposition flux. A uniform bed was assumed. Accordingly, a constant value for the critical stress for erosion of 0.1 N/m^2 was selected since it gave the best fitting to the measured bed profile. This value felt withing the range defined by Winterwerp and Kesteren (2004) and by Lumborg (2005) as was elaborated in Sub-section 4.4.2.

In Figure 7.2 the predicted bed profile shows a reliable agreement with the measured profile at the end of September 2012. Table 7.2 presents the parameters that were used in the calibration. The standard deviations of the measured water levels and bed levels were respectively 0.3 and 0.4 m. The RMSE is less than half the standard deviation which indicates good results. The high values of EF and the low values of ME and MAE confirm that the model can be used in the simulation of sediment transport. Figure 7.3 illustrates the sediment concentrations at the end of the first reach during the study period.

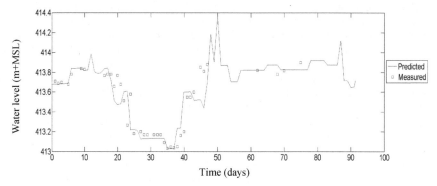

Figure 7.1. Measured and predicted water level downstream of Zananda offtake

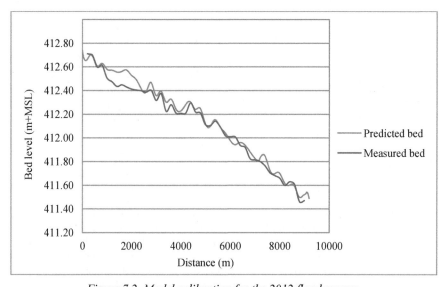

Figure 7.2. Model calibration for the 2012 flood season

Table 7.2. Parameters used in the calibration

Parameter description	Water level D/S the offtake	Bed profile values
MSRE	0.02	0.02
ME (m)	0.23	0.36
EF	0.93	0.95
MAE (m)	0.054	0.05

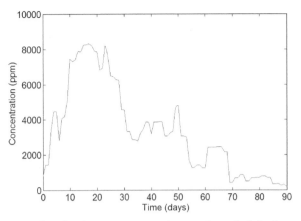

Figure 7.3. Predicted sediment concentrations at the end of the first reach (K 9.1)

7.1.3 Model validation

After the calibration the model was checked to assure that it adequately performed the functions for which it was intended. That is to accurately estimate sediment transport in the canals by comparing the predicted sediment with the measured data in 2011. For validation the model was run with respect to the field data of 2011 as presented in Annex C as input variables (inflow, sediment concentration at the offtake and bed profile). The model setup was made with the crest levels of the cross structures (movable weirs) in the time series. The predicted bed profile and the measured profile at the end of the flood season in 2011 are presented in Figure 7.4. The total deposition in the first reach as predicted by the model was 9160 m³, while it was about 8570 m³ according to bathymetric survey. The model slightly overestimated the volume of the sediment deposition by about 7%, which is insignificant and may be due to the fact that consolidation was not considered in the model. The accuracy of the simulation was checked by using different parameters. The results are presented in Table 7.3. The standard deviation of the measured bed level was 0.26 m. The low values of ME, RMSE and MAE and the high value of EF indicate that the model can be used for further analysis. Figure 7.5 indicates that the model has the capability to predict the sediment concentrations at any location during the simulation period.

Table 7.3. Result of the validation by using different parameters

Parameter description	Value
MSRE	0.023
ME (m)	0.29
EF	0.94
MAE (m)	0.068

7.1.4 Model verification

The results of hydrodynamic computation (water surface profile) were compared with the results of the DUFLOW model. DUFLOW is a micro-computer program for simulating one-dimensional steady/unsteady flow in open canals (Stowa and MX. Systems, 2002). In Figure 7.6 DUFLOW model schematization of Zananda Major

Canal is shown. The water surface profile was predicted by using DUFLOW and FSEDT models. The models were applied to Zananda Major Canal. The model setup was based on the geometric data of the canal as presented in Table 7.4 and on the following data:

- two weirs with crest levels of 411.03 m+MSL and 410.6 m+MSL located at 9.1 km and 12.5 km from the offtake;
- inflow at the upstream boundary is 5.5 m³/s;
- lateral flows at 7.2, 9.1 and 12.3 km from the offtake are 0.9, 0.92 and 1.01 m³/s;
- the water level at the downstream (tail of the canal) is 409.88 m+MSL.

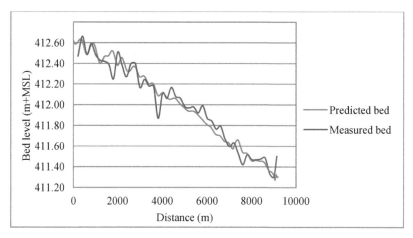

Figure 7.4. Model validation

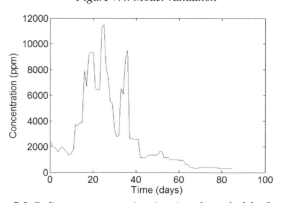

Figure 7.5. Sediment concentration (ppm) at the end of the first reach (K 9.1)

Table 7.4. Geometric data of Zananda Major Canal

Characteristics of the canal	Reach-1	Reach-2
Length of the reach (m)	9200	3200
Bed slope (-)	0.00013	0.00018
Bed width (m)	6	5
Width of the weir at the end of reach (m)	2	1.3
Side slope (-)	1:1	1:1

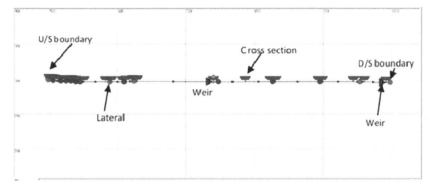

Figure 7.6 DUFLOW model schematization of Zananda Major Canal

In the DUFLOW model the general equation for free flow over a weir for a rectangular cross-section is based on Equation (3.10) as follows (Stowa and MX. Systems, 2002):

$$Q = 1.7wh^{1.5} \tag{7.5}$$

Where:
h   = water depth over the sill (m)
w   = width of the crest (m)

Figure 7.7 displays the difference between the water surface profiles computed by the two models, the variation is between 2 cm and 3 cm.

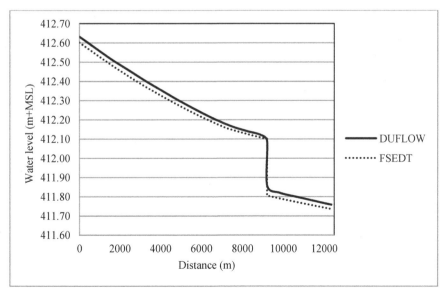

Figure 7.7. Water surface profile at Zananda Major Canal by using DUFLOW and FSEDT models

The hydrodynamic part of the model was functioning adequately since the difference between the predicted water surface profiles by the two models was reasonable.

7.2   Sensitivity analysis

Sensitivity analysis is a technique used to determine how different values of an independent variables (roughness coefficient, settling velocity, critical shear stress of erosion, longitudinal coefficient) affect sedimentation (dependent variable) under a given set of assumptions. In other words, the sensitivity analysis was performed in order to identify the effects of various inputs on the model's output. This analysis stresses the need for more and better field observations and greater accuracy to specify the input (physical properties), and indicates the most important physical properties.

The cross structures in the major canal were originally designed to be fully open by setting the crest level at its lowest position. However, sometimes they were adjusted at different positions. The justification behind the changes of the settings by the operators of the canals was to prevent canal breakages during periods of excess rainfall. The effect of changes in the crest setting on the sediment deposition has been investigated with M = 0.0016 kg/m^2/s and n = 0.023 s/m$^{1/3}$. Three options of gate settings have been addressed: the crest level at the lowest position and rise of the crest by 15% and 35%. The corresponding depositions for each option were 9,450, 9,860 and 10,600 m^3 respectively for the first reach and 5350, 6030 and 6610 m^3 respectively for the second reach. Figure 7.8 shows that the rise of the crest level leads to more accumulation of sediment in front of the cross structure and its effect extends to about 3 km upstream. In order to condense the accumulation of the sediment, the crest level needs to be set at the lowest position to let the sediment pass to downstream.

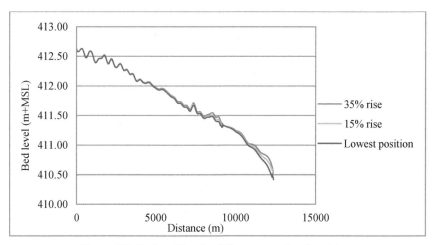

Figure 7.8. Bed profiles for different crest level positions

A calibration factor (k) was introduced in Equation (4.2) with settling velocity to address the effect of different fractions on the sediment transport. The deposited volumes when k = 1, 1.5 and 2 were 7030, 10,200 and 13,200 m^3 respectively for the first reach and 4150, 5530 and 6660 m^3 for the second reach. The effects of the different k values on the sediment deposition are depicted in Figure 7.9. More sediment was deposited at the head and it reduced towards the downstream. The model showed a high sensitivity to the calibration factor, since the increase in the calibration factor of settling velocity by 10% provided an increase in the volume of deposition by 8.8% for the first reach and 6.1% for the second reach.

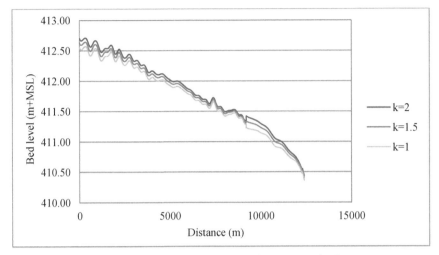

Figure 7.9. Effect of different sediment fractions in the deposition

The sensitivity of the bed roughness to the amount of sediment deposition was tested. The effect of changing the Manning coefficient from 0.025 to 0.03 is shown in Figure 7.10. Increase of the Manning coefficient by 20% slightly increased the deposition by 11.2% for the first reach and 10.8% for the second reach.

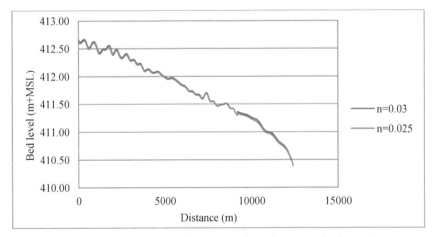

Figure 7.10. Effect of the change in bed roughness on the deposition

The sensitivity of changes in the critical shear stress of erosion was checked. When the critical shear stress for erosion was set at 0.06, 0.075 and 0.1 N/m^2 the corresponding deposited values were 8340, 9440 and 10200 m^3 for the first reach and 4040, 4720 and 5550 m^3 for the second reach. Figure 7.11 shows the predicted bed profiles according to the various critical shear stresses for deposition.

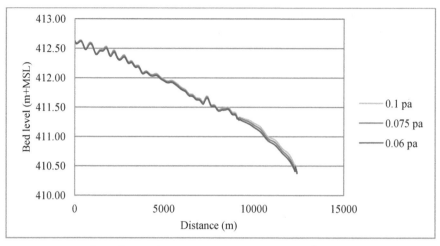

Figure 7.11. Effect of critical shear stress for erosion in N/m² (pa) in the deposition

However, the change in the erosion rate has no significant effect on the sediment deposition. An increase of the erosion rate by 10% decreased the deposition by only 0.2% for first reach and 0.8% for second reach. The erosion rates when M = 0.001 kg/m²/s were 10,200 and 5530 m³ for the first and second reaches respectively. When M = 0.002 kg/m²/s the deposition results were 10200 and 5070 m³ for the first and second reaches respectively. Therefore, the erosion flux depends mainly on the shear stress and the critical shear stress for erosion. The model demonstrated that changes in the settling velocity have a significant effect on the sedimentation in irrigation canals. The second reach showed a higher sensitivity to changes in critical shear stress and gate setting than the first reach. The effects of the various parameters on the sediment deposition in Zananda Major Canal are summarized in Table 7.5

Table 7.5. Summarized effect of the various parameters on the sediment deposition in Zananda Major Canal

Change of parameters by 10%	Increase of deposition in 1st reach (%)	Increase of deposition in 2nd reach (%)
Critical shear stress of erosion (τ_c)	4.3	6.5
Bed roughness (n)	5.6	5.4
Change of crest levels	3.1	7.6
Settling velocity (w_s)	8.8	6.1
Erosion rate (M)	0.2	0.8

7.3   Simulation of different operation scenarios of the major canal under study

The sediment transport in the canals under study was simulated by using the developed model FSEDT. The behaviour of the sediment and water delivery practices were investigated at the first and second reaches of Zananda Major Canal. The current operation system was evaluated and two options of operation were tested regarding the sedimentation. The different scenarios were compared under the existing conditions based on the data collected during the flood season in 2011. The sediment has also been simulated under future changed conditions and compared with the current operation.

7.3.1   Suspended sediment dynamics

The study of the dynamic behaviour of the sediment in the irrigation canal network under varying flow conditions is a prerequisite for attaining efficient system operations. The dynamics of sediment along the first reach of the Zananda Major Canal have been investigated for a 3 days simulation period with a constant sediment concentration of 6000 ppm. It is obvious that the sediment concentration reduces towards the downstream as a result of reduction of the sediment load, which leads to less deposition. Although the velocity is reduced towards the downstream as was shown in the Tables 5.6 and 5.7 it was recognized that the deposition reduced towards the downstream as well. At low shear stress (less than the critical shear stress for erosion that was given by 0.1 N/m^2) the amount of deposition does not depend upon the flow condition. The governing factors are the concentration and settling velocity. However, the increase of the shear stress above the critical value for erosion led to more re-suspension and less deposition downstream. It should be noted that the increase of the shear stress towards the downstream is due to the effect of the increase in the energy slope, which is more than the effect of reduction in the water depth on the shear stress (water surface profile is M2 curve). Furthermore, more water supply to the system means a high sediment load and a high deposition rate. The deposition values were 107, 157 and 179 m^3 when the inflow was 1.75, 3.5 and 5.5 m^3/s respectively. Figure 7.12 shows the effect of the different flow conditions on the sediment load, sediment concentration, shear stress and deposition along Zananda Major Canal after the 3 days simulation period.

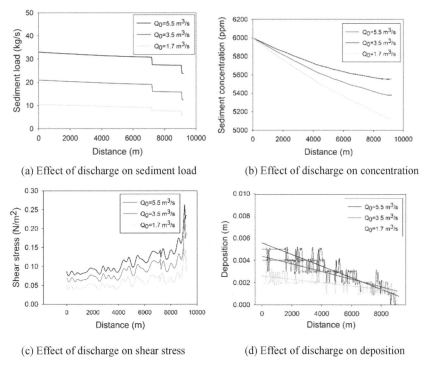

(a) Effect of discharge on sediment load   (b) Effect of discharge on concentration

(c) Effect of discharge on shear stress   (d) Effect of discharge on deposition

Figure 7.12. Addressing of the sediment processes along the canal under different flow conditions

7.3.2   Operation based on data collected in 2011

The role of the district engineer in Gezira Scheme is to distribute the water to the canals and fields according to the water duty as specified by the agriculture inspectors. It was found that the actual water flow delivered to the system was more than the required quantity in most of the cases.

The data that were applied to the model are presented in Annex C. It was observed that the crest levels of the cross structures were set at their lowest position at 411.10 m+MSL and 410.35 m+MSL for most of the time for Structure-1 and Structure-2 respectively. Therefore, the crest levels were assumed constant along the simulation period. The bed silted up above the design bed level with 1.26 m in front of the weir and 0.9 m at Structure-1 and Structure-2. This reduces the water depth at the downstream boundary. The water depth is equivalent to the height of the crest above the bed level and the upstream water depth over the crest that was obtained from the stage discharge relationship of the control structure. As a result of sedimentation the height of the crest was reduced. Therefore, the water depth along the canal was consequently reduced. A total deposition of 9160 m³ was predicted by the model for the first reach. Figure 7.13 illustrates the predicted bed profile and shows more deposition at the head of the reach.

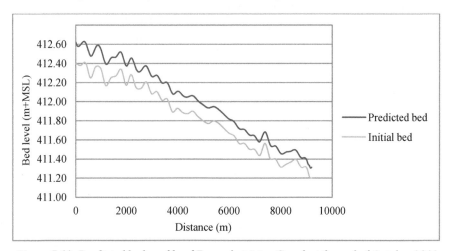

Figure 7.13. Predicted bed profile of Zananda Major Canal at the end of October 2011

7.3.3   Scenario I: operation based on the indent system (authorized operation)

The simulation for Scenario I was performed to investigate the sediment behaviour when the canal is operated based on the indent system. The water duty is a procedure for water allocation that was adopted in Gezira Scheme as was elaborated in Section 3.2. Although this system (indent) was absent during the last years it can give an approximate estimation of the water requirement over a season. The duty for the study area was estimated at 71.4 m³/ha/day (30 m³/feddan/day) for all crops, including field losses at the head of the Abu Ishreen canal (Plusquellec, 1999). The peak of the CWR for different crops is around this value (Adam, 2005). The delivered discharge to Zananda Irrigation System based on the duty and the actual irrigated area in 2011 is presented in Table 7.6.

The sowing date for cotton was between 1 and 10 August, for groundnut it was between 20 and 30 July and for sorghum between 1 and 10 June. Therefore, the indent during July was based on the cropped area with sorghum (excluding the first irrigation), groundnut and garden. From 1 August, cotton was added to the cropping pattern. The results of the simulation of this system show that there was a reduction in the deposition by 3150 m^3 for the first reach and by 1750 m^3 for the second reach. Compared to the actual situation the reductions were 34 and 40% for the first and second reach respectively. Figure 7.14 shows the bed predicted according to the different operation scenarios.

Table 7.6. Irrigated area in 2011 and discharge based on the duty of 71.4 m^3/ha/day

Canals	1 - 30 July		1 August - 30 October	
	Area (ha)	Indent (m^3/s)	Area (ha)	Indent (m^3/s)
Gimeliya	214.0	0.18	302	0.25
G/ Hosh	306	0.25	450	0.37
Ballola	213	0.18	228	0.19
W/ Elmahi	239	0.20	322	0.27
G/ Abu Gimri	92.4	0.08	157	0.13
Toman	395	0.33	460	0.38
Gommuiya Br	391	0.32	555	0.46
Total area of Zananda	2840	2.34	3800	3.14

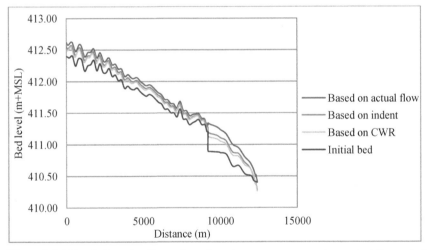

Figure 7.14 Simulation of sediment based on the crop water requirement at the period of high sediment concentration, the predicted bed based on the indent system and based on the actual flow

The model was set with the design profile and canal cross-sections, while the input data remained the same as in the above scenario. The crest levels of the cross structures were set at their lowest position. The sediment concentration at the offtake of the canal in 2011 was set as upstream boundary condition. The predicted sediment deposition was 11,400 m^3 the first reach and 4540 m^3 for the second reach. Compared to the actual situation the results show an increase in deposition of about 24% and 4% at the first and the second reach respectively. The increase in sediment deposition was due to the shear stress that was less than the critical shear stress for erosion as a result of changes in the canal slope. The design slope was 10 cm/km for the first reach and 5 cm/km for the second reach. The design bed widths were 6 and 5 m for the first and second reaches respectively. In fact, the model predicted more sediment deposition per km in the second reach when compared to the first reach. This is attributed to the reduction in the slope from 18 cm/km to 5 cm/km as shown in Figure 7.15.

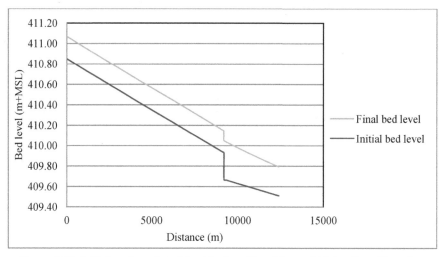

Figure 7.15. Initial and predicted final bed profile of Zanand Major Canal based on duty and design conditions

7.3.4 Scenario II: operation based on the crop water requirement

The aim of Scenario II was to mitigate the inflow of sediment during the high concentration period and to address its effect on the deposition process. The inflow at the offtake was computed based on the crop water requirement by considering all the losses as given in Table 7.7 and already discussed in Sub-section 5.4.1. The sediment simulation was based on the crop water requirement during the period of high sediment concentration between 10 July and 10 August 2011, since there was more sediment inflow to the scheme during this period. The inflow during that period reduced from 11.5 Mm^3 to 4.86 Mm^3 based on CWR (57.7% reduction in the inflow). Consequently, compared to the actual flow the sediment volume was reduced by 51% in the first reach and by 55% in the second reach. The predicted bed profile compared to others is also shown in Figure 7.14.

7.3.5 Operation under future conditions

A hypothetical case has been analysed to predict the deposition in Gezira Scheme in case of a reduction in the sediment concentration as a result of the construction of the Grand Ethiopia Renaissance Dam. This is a dam under construction by the Ethiopian government across the Blue Nile River near the boundary of Sudan, upstream Roseires Dam. The sediment concentration is expected to reduce as a result of this dam according to Ali (2014). The amount of reduction will mainly depend on the operation policy. The aim of this scenario was to address the possible effect on the sedimentation in Gezira Scheme if the sediment concentration of the Blue Nile River would be reduced by 50%. The results of the simulation of the suspended sediment transport in Zananda Major

Canal indicate that the deposition in the first and second reach may be 2410 m³ in the first reach and 832 m³ in the second reach. Compared to the actual situation in 2011 this implies respectively 74 and 81% reduction in the deposition. The predicted bed level during the simulation period as presented in Figure 7.16 shows that there is a sharp increment between day 15 and day 40 (15 July - 10 August), which indicates that the deposition occurred within a short period. Table 7.8 summarizes the siltation in Zananda Major Canal according to the different options of operation.

Table 7.7. Release based on the crop water requirement (Mm³) in 2011

Period	Zananda	Gemilia	G/Elhosh	Ballola	W/ Elmahi	Toman	Gemoia	G/Abu gomri
Jul I	3.93	0.34	0.36	0.34	0.38	0.64	0.37	0.10
Jul II	1.19	0.08	0.13	0.09	0.10	0.16	0.14	0.03
Jul III	1.32	0.09	0.15	0.10	0.11	0.18	0.15	0.03
Aug I	2.04	0.17	0.25	0.11	0.18	0.23	0.24	0.06
Aug II	1.94	0.16	0.21	0.13	0.17	0.25	0.22	0.05
Aug III	2.46	0.18	0.24	0.15	0.20	0.30	0.25	0.06
Sept I	2.44	0.20	0.27	0.16	0.21	0.32	0.28	0.07
Sept II	2.56	0.21	0.29	0.16	0.21	0.33	0.30	0.07
Sept III	2.68	0.21	0.31	0.17	0.22	0.34	0.32	0.07
Oct I	2.64	0.21	0.31	0.16	0.22	0.32	0.31	0.07
Oct II	2.44	0.20	0.29	0.14	0.21	0.28	0.29	0.07
Oct III	2.28	0.18	0.29	0.12	0.19	0.24	0.28	0.06

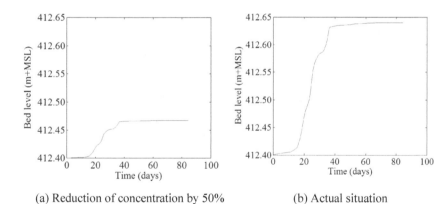

(a) Reduction of concentration by 50%   (b) Actual situation

Figure 7.16 Accumulation of the sediment at the offtake of the major canal based on reduction of concentration by 50% and on the actual situation

Table 7.8. Canal siltation in (m³) based on different options of operation by using 2011 data

Operation scenarios	1st reach	2nd reach
Actual flow	9160	4370
Indent system	6010	2620
Apply CWR during high concentration period	4510	1970
50% reduction in concentration	2410	832
Design bed profile with indent system	11400	4540

7.4   Operation of the minor canals under different operation scenarios

7.4.1   Operation under the night storage system

The aim of the scenario on the operation under the night storage system was to get an answer to the question what will happen regarding the sediment transport if the night storage is still in use. The FSEDT model was developed to simulate continues steady state flow. Thus, it was not practical to apply the model for the night storage scenario practically at the filling period. The flow at night is typically unsteady flow, since the discharge, water depth and flow velocity change with time. The absence of historical records when the night storage system was in use make the calibration and validation of this scenario rather difficult. Therefore the Manning roughness was set 0.025 s/m$^{1/3}$ as in the design. It was assumed that the critical shear stress of erosion does not change since the factors, which affect erodability such as the physical, physico-chemical and biological properties of the sediment do not change.

The DUFLOW model was applied to simulate the hydrodynamic part at night (filling time). The unsteady flow is characterised by the change of the flow rate with time and location as in the case of wave propagation. The solution is found by solving the continuity equation and momentum equation (Saint Venant equation) by using the Preissmann scheme. The output of the model was the water depth and discharge that was used as input to simulate the suspended sediment transport.

The computation of the sedimentation was carried out by using a spread sheet. For the night storage system the upstream brick wall of the weir, on each side of the frame of the sluice gate was cut away to the full supply level (Farbrother, 1974). When the gates are shut at sundown the reach fills up to overflow the weir formed by the remaining brickwork. Figure 7.17 shows the night storage weir. The weir was treated as broad crested weir since the upstream head is smaller related to the top length of the weir. The weir is 3 m wide, 0.3 m long and the maximum upstream water depth above the crest is 0.2 m. See Sub-section 3.1.5.

Figure 7.17. Night storage weir before and during the flood season

Hydrodynamic simulation by using the DUFLOW model

The schematization of Toman Minor Canal was set in the DUFLOW model as shown in Figure 7.18. The DUFLOW model has the capability to predict the change in the water level and storage time when the reach achieves its equilibrium condition (the inflow equals to the outflow). The canal is divided in 4 reaches, with a cross-sectional area as defined at the beginning and the end of each reach. The characteristics of the Toman Minor Canal are presented in Table 7.9. The design bed profile, the length of each reach, the bed levels at the beginning and end of each reach and the characteristic of the cross structures were defined for the model. The initial water levels at each reach were set at the full supply level (FSL). The crest of the first and second night storage weirs were 410.75 m+MSL and 410.52 m+MSL respectively. The pipe regulator was treated as orifice with neglected friction losses, since it has a short length (diameter 0.76 m and length about 2 m). The inflow was 0.46 m³/s and defined as the upstream boundary condition. The water level at the tail of the system was 410.75 m+MSL and was defined as downstream boundary condition. Then, the model was run for one day with time steps of 5 minutes.

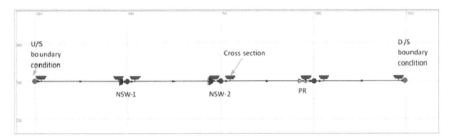

Figure 7.18. Schematization of Toman Minor Canal by using the DUFLOW model

Table 7.9. Geometry of Toman Minor Canal

Reach	Bed width (m)	Side slope (-)	Bed slope (-)	Length of the reach (m)
1	2	2	0.0002	1200
2	2	2	0.0002	1800
3	1.5	2	0.0001	1700
4	1	2	0.00025	1300

The model comes up with the final water profile (night storage level (NSL)) and the filling time for each reach. Different inflows of 0.25, 0.32 and 0.46 m³/s at the offtake of Toman Minor Canal were tested. The corresponding increments in the water level (NSL) above the full supply level (crest level) at the night storage weirs were 0.13, 0.16 and 0.2 m respectively. For the different discharges the storage times of each reach are presented in Table 7.10. Therefore, it was concluded that the only case that was in agreement with the night storage according to the design, was the case where 0.46 m³/s supplied to the canal (operation of the canal at full capacity), which takes a filling time of about 12 hours and a water depth over the crest level of the night weirs was 0.2 m. Based on this, the flow at the head of the canal was kept constant at the design full supply value 0.46 m³/s. The field canals can only take their fair share of water if the minor canal runs at full supply.

The DUFLOW model shows how the water level and discharge change with time during the night from 6:00 pm to 6:00 am when the canal is operated at the full capacity of 0.46 m³/s. Figure 7.19 displays the change in water levels just upstream of the first and second night storage, and the pipe regulator. The first reach takes less time whereas the second reach takes more time. The third and fourth reaches achieved stability at the same time due to the existence of the pipe regulator as control structure between the two reaches as shown in Table 7.10. Figure 7.20 demonstrates that there was no storage detected in the third reach, since the NSL was approximately at the same level of FSL. The change in inflow for each reach during the filling time is shown in

Figure 7.21. The outflow equals the inflow after 5, 9.5 and 12 hours at the first, second and third-fourth reach respectively. The results show that the stability of the system starts from upstream towards downstream.

Table 7.10. Filling time of each reach for different inflow

Reach	Time needed for the reach to be filled in hours		
	$Q = 0.46$ m³/s	$Q = 0.32$ m³/s	$Q = 0.25$ m³/s
1	5	5.5	6
2	9.5	9.17	8
3	12	9	7.75
4	12	9	7.75

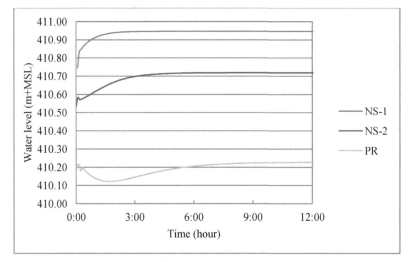

Figure 7.19. Change in the water level upstream of the cross structures; night storage (NS-1 and NS-2) and pipe regulator (PR) for $Q = 0.46$ m³/s

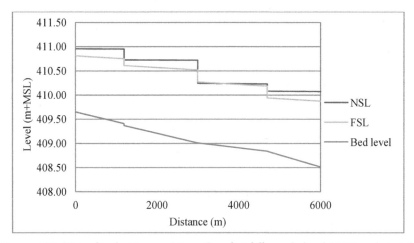

Figure 7.20. Water level at Toman Minor Canal at full supply level (FSL) and at night storage level (NSL) for $Q = 0.46$ m³/s

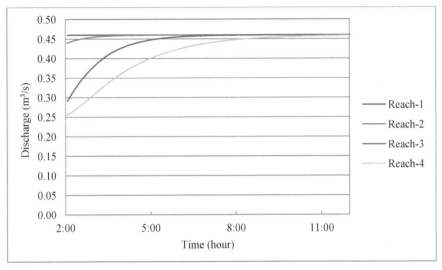

Figure 7.21. Inflow at the head of each reach in 12 hours

The simulation of the sediment under the night storage operation is divided into two parts:

- sedimentation during the filling time from 6:00 pm to 6:00 am;
- sedimentation during the irrigation period from 6:00 am to 6:00 pm.

Sediment simulation during the filling time

The output of the hydro-dynamic computation by using DUFLOW was used in the computation of the suspended sediment transport. A spread sheet was developed to compute the sediment every one hour. The same concept of the sediment computation by the FSEDT model for the first time step was followed (Section 6.2). The steps of the computation are as follows:

1.   the water depth and discharge at the computation points (every 400 m) are obtained from the DUFLOW model;
2.   the constant inflow of 0.46 m³/s (full capacity) and the sediment concentration of 6000 ppm are defined in the model as upstream boundary condition. Thus the inflow sediment load is known;
3.   the shear stress is computed at each computation point based on Equations (6.25) and (6.32). The results indicate that the flow induced bed shear stress is less than the critical value of 0.1 N/m² as that used in the model based on the calibration (re-suspension (E) = 0).
4.   the deposition is calculated based on the approach as described in Section 6.2 at t_0. Equation (4.2) is used to obtain the sediment deposition rate. The settling velocity is obtained from the correlation between the settling velocity and the concentration in Equation (5.1) ;
5.   sediment budget approach is applied and the sediment load is computed at the computation points from Equation (6.56). Then the concentration is obtained since the discharge is known at each point;
6.   all the above steps are repeated till the end of the reach. Then the outflow sediment load from each reach is the sediment inflow to the next reach, since no deposition is expected at the control structures.

The distance between the computation points (400 m) was divided into 46 m distance steps as adopted by the FSEDT model. The deposition weight was converted into volume for a sediment density of 1200 kg/m³. The results of this scenario are presented in Annex D. The results indicate that more deposition occurs during the first hours, since coarser material settles faster. Then, the rate of deposition reduces and approximately remains constant as shown in Figure 7.22. The cumulative deposition during 12 hours is presented in Figure 7.23.

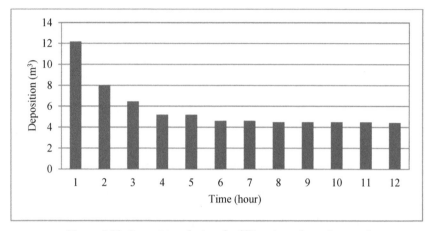

Figure 7.22. Deposition during the filling time along the canal

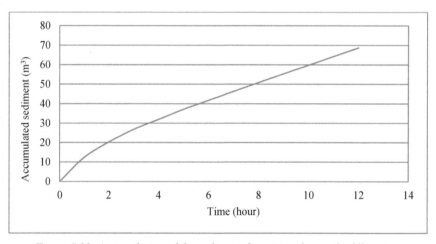

Figure 7.23. Accumulation of the sediment deposition during the filling time

Sediment simulation during the irrigation period

The field outlet pipes are opened at daytime to release the water from the minor canal to the field. The water level reduces gradually during the daytime from NSL till they reach the FSL at 6:00 pm. FSEDT model has been applied to simulate the sedimentation during the daytime. The model was set with the constant discharge of 0.46 m³/s, concentration of 6000 ppm and the design bed profile. The model was run for 12 hours to predict the deposition. The total amount of deposition (filling period and irrigation period) along Toman Minor Canal is presented in Figure 7.24. More sediment deposition was detected in the second reach. This was due to the fact that here the storage volume is large as a result of its long length (the longest reach is 1800 m) and the height of the break wall exceeds that of the first reach by 17 cm. The deposition during the night was 60, 77, 81 and 78% of the total deposition for the first, second, third and fourth reaches respectively. It was concluded that the average deposition during the filling time is about 74% of the total volume of deposited sediment, which coincides with the water column test results (Sub-section 5.6.2), although the concentration is different.

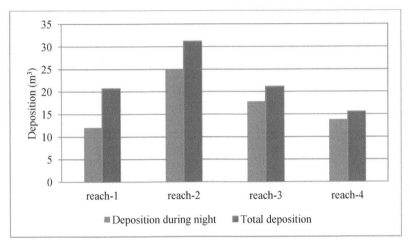

Figure 7.24. Sedimentation along Toman Minor Canal for one day based on the night storage system

Continuous operation system versus the night storage system

The sediment transport was simulated based on continuous operation by applying the FSEDT model. The FSEDT model was set with similar input data used to simulate the sediment during the daytime (irrigation period) in the night storage system. In continuous operation, the amount of deposition reduced towards the downstream. The results of the simulation of the sediment for one day are summarized in Table 7.11. It was found that the continuous system reduces the sediment deposition by 55% when compared with night storage system.

Table 7.11. Comparison between the amounts of deposition during night storage and continuous operation for one day

Reach	Deposition (m^3)			
	Night storage system (NSS)		Continuous system	Reduction in deposition (%)
	During night (12 hours)	Total (24 hours)	Operation (24 hours)	
1	12.1	20.7	17.2	17
2	24.9	31.2	12.5	60
3	17.8	21.1	6.6	69
4	13.8	15.5	3.5	77
Total	68.6	88.5	39.8	55

7.4.2 Operation under the actual discharge

The current condition of the minor canal is far from the design (appropriate situation). The sediment has closed more than 50% of the gates of the cross structures (night storage system) as was shown in the right photo in Figure 7.17. The water levels are most of the time above the sill of the weirs as a result of bed rise. Therefore, part of the water passes through the sluice gates and the rest passes over the weirs. Farmers use sacks to control the water passing to downstream. It is therefore not easy to detect the gate settings. It was observed the NS-1 was continuously overtopped, whereas at NS-2 the overtopping occurred occasionally due to the higher level of the crest compared to the first one. The pipe regulator at the last reach was fully corroded and the water bypassed underneath the bottom of the pipe to the last reach as displayed in Figure 7.25. Furthermore, the field out let pipe (FOP) was far below the bed level of the canal. Farmers always excavate in front of the FOP to let the water pass to the field canals. Hence a pool has been created. In addition to all that, the command is very low at some parts. The irrigation schedule is unattended to and no irrigation schedule is followed as was explained in Sub-section 5.4.2.

Figure 7.25. Pipe regulator at the head of the last reach

Figure 7.26 shows that the size of Abu Ishreen canals varies from one to another. Albers (2012) studied the water management in Toman Minor Canal in 2011 and stated that there is no clear pattern of the cross-section profile over the reach due to sedimentation and weed growth. The study also indicated that the average discharge of Abu Ishrean per reach is varying between 0.04 m³/s in the first reach and 0.032 m³/s in the last reach. It was recognized that cropping intensity can be less than 50% and part of the Nimras is not cropped.

All the above mentioned reasons make the simulation of the actual situation in Toman Minor Canal very difficult. Thus the model was setup with the design canal profile, the inflow and sediment concentration at the upstream boundary based on collected data in 2011 and 2012. Based on the flow measurement and the canal geometry it can be stated that the roughness is varying from reach to reach. The sections of the first and the fourth reach are rather parabolic than trapezoid. The values of the roughness are shown in Table 7.12. Based on the field measurements irrigation rotation is applied in the model with varying discharge of the Abu Ishreen canals between 0.02 m³/s and 0.12 m³/s according to field measurement.

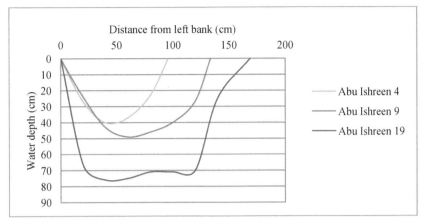

Figure 7.26. Measured cross-sections of the flow for selected field canals on 29 September 2011

Table 7.12. Manning roughness along Toman Minor Canal

Location	Discharge (m³/s)	Bed width (m)	Side slope	Water depth (m)	Bed slope	Surface width (m)	n (s/m^{1/3})
reach-1	0.4	Parabolic	-	0.6	0.00017	2.8	0.019
reach-2	0.32	1.8	1.6	0.43	0.0002	3.18	0.022
reach-3	0.14	1.8	1.6	0.37	0.00018	2.98	0.036
reach-4	0.22	parabolic	-	0.6	0.0002	4.5	0.036

In 2012 the offtake of Toman Minor Canal was closed between 12 July and 1 September. A small amount of water (seepage) passed to the canal over the fixed beam (between 12 July and 26 July) due to the high water level upstream of the head structure. The suspended sediment transport was simulated and the predicted sediment deposition was about 1430 m³. The distribution of the sediment is illustrated in Table 7.13. The result shows that the deposition reduced towards the downstream. However, HR Wallingford (1990a) reported that 60% of the deposition occurred in the first reach, while in this study the average deposition was 34% in the first reach. The difference is referred to change in the canal geometry and flow conditions. The deposition represented 37% of the inflowing sediment load of 3.55×10^3 m³ (4.27×10^3 ton). The bathymetric survey illustrated that the average deposition during the flood season in 2012 was about 20 cm. The volume of deposition 1320 m³ coincides with the volume obtained by the model. Compare to the survey the model shows reasonable results although the canal geometry is different.

In 2011 the offtake was closed for one month as a result of rainfall on 8 August after the period of high sediment concentration as shown in Figure 7.27. The high water level recorded on 4 July 2012 and the discharge of 0.68 m³/s exceeded the full capacity. This was due to the fact that a too high water level was kept upstream of the canal inlet while the crest level was set at its lowest position. The sediment delivered to the canal during the study period in 2011 was about 6,180 m³. However, the predicted deposition by the model was 2,360 m³ as shown in Table 7.13. The deposition represents about 38% of the inflowing sediment load.

The bed profile at the end of the flood season in 2011 was not measured since unexpected canal clearance activities were conducted in September. Figure 7.28

presents the predicted bed profile with respect to the actual flow conditions and sediment properties in 2011 and 2012. More deposition was detected in 2011 as a result of the high sediment load released into the canal during the study period. The accumulated sediment at the head of the first reach is plotted in Figure 7.29. It shows a bed rise of about 25 cm in 2011 and 15 cm in 2012.

Table 7.13. The predicted amount of sediment deposition along Toman Minor Canal

Location	2011		2012	
	(m^3)	%	(m^3)	%
reach-1	820	35	455	32
reach-2	733	31	440	31
reach-3	526	22	342	24
reach-4	281	12	193	13
Total	2360		1430	

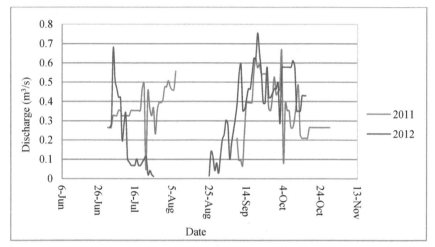

Figure 7.27. Inflow in Toman Minor Canal between July and October 2011 and 2012

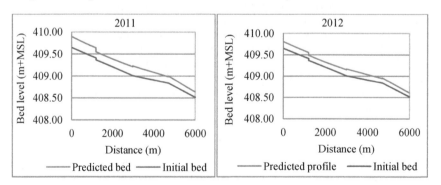

Figure 7.28. Simulation of the sediment in Toman Minor Canal

7.4.3   Operation under the crop water requirement

Suspended sediment transport was simulated according to the crop water requirement between 10 June and 10 August. Table 7.14 presents the crop water requirement and actual flow during the flood season in 2011. The implication of the reduction of the water delivery during the period of high sediment concentration was tested. The model was setup with the design bed profile, sediment concentration and inflow of 2011. The depositions were 623, 526, 361 and 176 m^3 for the 1st, 2nd, 3rd and 4th reach respectively. When matched with depositions that were obtained based on the actual flow in 2011 a reduction of 29% was detected.

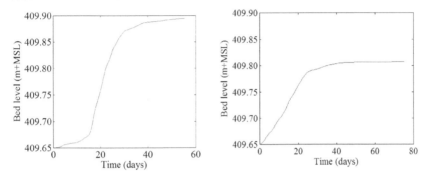

Figure 7.29. Accumulation of the sediment at the head of the first reach in 2011 and 2012

Table 7.14. Release based on the crop water requirement (CWR) and actual flow in Toman Minor Canal in 2011

Period	CWR (Mm^3)	Actual flow (Mm^3)
JUL-I	0.64	0.27
JUL-II	0.16	0.31
JUL-III	0.18	0.30
AUG-I	0.23	0.29
AUG-II	0.25	Closed
AUG-III	0.30	Closed
SEPT-I	0.32	Closed
SEPT-II	0.33	0.32
SEPT-III	0.34	0.42
OCT-I	0.32	0.31
OCT-II	0.28	0.24
OCT-III	0.24	0.21

7.4.4   Operation under the design discharge with different field outlet pipe capacities

While the irrigation system in Gezira Scheme was designed based on the demand system the users could obtain the water directly from a supply point (field outlet pipe) with 7 days fixed duration. Nowadays, the field outlet pipes are below the bed level of the minor canals as a result of bed rise and the users are facing the flow blockage problem due to the piling up of sediment. However, as a result of crop intensification, sedimentation and poor maintenance, the field canals are unable to deliver the design discharge of 0.115 m^3/s to the farms. The aim of the scenario based on the design discharge with different field outlet pipe capacities was to address the impact of reducing the FOP capacity in conjunction with sedimentation in the minor canals. Three options have been investigated; when the FOP is at full capacity, at 75% and at 50% of the full capacity.

In Gezira Scheme; 0.42 ha (one feddan) needs 400 m^3 of water to be irrigated during each irrigation interval. The Abu Ishreen canals were designed to irrigate 37.8 ha (Nimra) in 7 days with 14 days irrigation interval if operated daily for 12 hours at full capacity (Ahmed et al., 1989). Based on this information the durations of different operation schedules were anticipated and the sediment deposition was predicted when operating the canal continuously. The model was setup with the canal at full capacity of 0.46 m^3/s and a Manning coefficient of 0.022 (according to the design). The sediment concentrations of 2011 were introduced to the model as upstream boundary condition. A rotational schedule has applied. Three options of average rates of the water flow through the field canals (Abu Ishreen canals) were applied:

I.   operating Abu Ishreen at full capacity (0.115 m^3/s) with 16 days irrigation interval and 4 days irrigation duration;
II.   operating Abu Ishreen at 75% of the full capacity (0.086 m^3/s) with 15 days irrigation interval and 5 days irrigation duration;
III.   operating Abu Ishreen at 50% of the full capacity (0.058 m^3/s) with 14 days irrigation intervals and 7 days irrigation duration.

Different numbers of FOPs were opened per day to deliver the water to the fields. Three field canals were operated per day based on option-I, four canals per day based on option-II and six based on option-III. The results of the deposited sediment along the canal are presented in Table 7.15. It was found that the difference in the amount of deposition is not significant along the minor canal, since the amount of water delivered to field canals was approximately the same for the different options.

Table 7.15. Deposition along Toman Minor Canal when different numbers of field pipes with different capacities were opened

Reach	Deposition (m^3)		
	Option-I	Option-II	Option-III
1	777	765	756
2	496	481	472
3	268	281	272
4	102	107	105
Total volume	1643	1634	1605

Furthermore, three Abu Ishreen canals were operated based on the above different options (I, II and III) of operation with irrigation intensities of 57, 48 and 28% respectively. The results as shown in Table 7.16 demonstrate a slight increment in the deposition with reduction of the capacity of the FOPs. When the capacity was reduced by 75 and 50% of the full capacity an increase in the deposition of respectively 22 and 31% was found. In the area supplied by Toman Minor Canal the crop intensity can be increased up to 57% when the FOPs are operated at full capacity. However, it is difficult to augment the intensity and reduce the water delivered to the field canals at the same time. This will increase the irrigation interval with the result that the plants suffer from water stress.

Table 7.16. Deposition along Toman Minor Canal when the same numbers of field
canals (Abu Ishreen) with different capacities were opened

Reach	Deposition (m^3)		
	Option-I	Option-II	Option-III
1	777	846	843
2	496	594	644
3	268	371	412
4	102	186	259
Total deposition	1643	1997	2158

8   Evaluation

In this chapter the most important findings of the study are presented. The output of the developed model was compared with previous studies and field measurements in Gezira Scheme. The effects of the sediment on the hydrodynamics of the canals are presented. The evaluation of the previous operation plans that were followed in Gezira Scheme have been investigated and compared to the current situation and to the proposed plans in terms of sediment and water management. Finally the conclusions are drawn and the proposed way forward is presented.

8.1   Effect of the sediment on the hydrodynamic behaviour of irrigation canals

Lack of funds for sediment management led to rising of the bed level of the canals far above the design level. The canal profiles indicate more sediment at the head of the canals and reduction towards the downstream. This confirmed the findings of HR Wallingford (1990a) and also coincides with the model results. As a result of bed rise the water levels became above the design level (FSL) and the water depth reduced. It was recorded at the inlet of Toman Minor Canal that the water level exceeded the beam level during 80% of the time of operation in 2012, which indicates that most of the time the structures are not functioning properly. The FSL cannot achieve the required water supply, which has resulted in inadequate water supply and inequitable water distribution. Accordingly, irrigation difficulties were detected at some parts of the scheme. In Gezira Scheme overtopping of the banks is perceived at some locations whereas water shortage has been detected in other parts. The reason behind poor performance of manually operated irrigation systems in terms of low equity of water distribution is the mismatch between the settings of the control structures and the water demand. Adjusting the water control structures at the head of the major canals and outlets in a random way and continued deposition have resulted in instability of the water level. Furthermore, lack of flow measurement facilities, communication systems and an inadequate number of operators have had a negative influence on the system.

In 2005 the farmers obtained the right to choose the cropping pattern. This has resulted in a variety of crops within the Nimra (plot area that is irrigated by a field canal) and in a cropping intensity within the Nimra that can be less than 50%. Managing the water for multi crops is not easy for farmers and operators. Consequently, this may be one of the reasons behind the absence of an irrigation schedule. For better water management, the Nimra would have to be planted with one crop type or crops that need more or less the same quantity of water and irrigation duration.

8.2   Sediment transport modelling

The challenge in modelling of cohesive sediments is that the sediment dynamics are described by a mixture of theoretical and empirical equations. The FSEDT model has been formulated by using Matlab software to simulate the fine suspended sediment transport. Different types of hydraulic structures are integrated in the model. Among them are adjustable overflow structures. The model can predict the bed profile, water level, sediment concentration and sediment load along a canal. It can also predict the concentration and deposition of sediment at any location during the simulated period.

The FSEDT model has been developed, calibrated and validated based on the data collected from field measurements in Gezira Scheme. The accuracy of the simulation has been tested by using different measures such as root mean square error, modelling efficiency, maximum error and mean absolute error. The input to the model is direct and can be simply introduced in the model.

There are a number of inherent limitations in the model. The equations are for one-dimensional flow. The flow of the water in a section is an average over the width and depth of that section. Furthermore, the model has been developed for one sediment fraction. The model overestimated the fine sediment deposition, since consolidation is not accounted for in the computation. The concern in this study was to quantify the sediment deposition as well as to simulate different operational plans at defined simulation periods for better sediment and water management.

8.3   Effect of different operation scenarios on the sedimentation

The focus in this study was on the simulation of the sediment during the flood season between July and October, since the sediment concentration after the flood season is much less than 200 ppm and can be neglected. The prediction of the bed profile between November and February shows no change, even not when a constant concentration of 200 ppm is applied if the canal flow is at full capacity. The analysis of the deposition process of cohesive sediment indicates that it is not easy for the sediment to re-suspend after the flood season, even not if the canal is operated at full capacity. However, improvement in the operation is needed during the flood season when most of the sediment deposits.

8.3.1   Simulation of the sediment in Zananda Major Canal

Although the sediment load and canal conditions are different from place to place within Gezira Scheme, the operation pattern is similar for all parts of the scheme. The sedimentation problem upstream of the irrigation system is significant when compared to downstream. Therefore, Zananda Major Canal has been selected since it is the first canal that receives water from Gezira Main Canal. The data collected for the canals under study were used in the calibration and validation of the model. Thus, the model can be applied at any irrigation sub-system within the scheme.

The simulation of the sediment for the situation in 2011 shows that as result of oversupply when less water is needed more deposition along the canals occurs. Two operation scenarios were tested; the indent system and operation based on the crop water requirement. The indent system of operation has been tested for the current and the design canal profile. The discharges at the head regulator and outlets have been set based on the duty. Less sediment has been predicted when the model was set with the actual canal geometry in 2011. The analysis shows a reduction of 34% in the first reach and of 40% in the second reach. In contrast, the deposition increased when the model was set at the design canal profile. This is due to the reduced velocity and boundary shear stress as result of reduction in slope and canal cross-section area. Thus, less re-suspension took place and the deposition increased correspondingly.

Simulating the sediment based on the crop water requirement during the period of high sediment concentration between 20 July and 10 August reduced the water supply in 2011 by 58% and the deposition by 51 and 55% at the first and second reach respectively. This option of operation shows less deposition compared with the indent operation system, since the indent system is based on a constant duty all over the season, which results in more water deliveries to the system (Plusquellec, 1999). Consequently, more sediment load and more deposition are expected.

Moreover, a hypothetical scenario has been generated to address the effect of the improvement of land use upstream in the Blue Nile River Basin and the impact of new dams on the sedimentation in Gezira Scheme. The reduction of the concentration by 50% results in a reduction of the deposition by 78% compared to the situation in 2011.

8.3.2   Simulation of the sediment in Toman Minor Canal

The night closure of irrigation canals has been investigated. The night storage system can improve the reliability of water supply by supplying more water during the daytime and increase in the command during the night. However, a higher deposition rate was detected in the minor canals. Simulation of the sediment transport revealed that about 74% of the sedimentation occurs at night. When the deposited volume is compared with the continuous irrigation system; the average increase in deposition is more than 122%. This kind of operation is not favourable, since it leads to more sediment deposition, causes overtopping and needs intensive maintenance to cope with this system.

During the period before intensification (mid-seventies), when the night storage system was applied, the 7 days irrigation duration usually satisfied the demand in a Nimra. After intensification it was difficult to apply the night storage system. Then the system was changed again into a continuous type of irrigation to fulfil the needs. In fact the night storage system cannot work properly if the crop intensity during the summer season exceeds 38%. When the field outlet pipes did not work at their full capacity as a result of sedimentation and lack of maintenance, the irrigation schedules have been changed accordingly and the duration of the irrigation has been increased. This may cause plant stress and affect the productivity. The reduction of the canal capacity by 75 and 50% increased the deposition in the minor canal by respectively about 22 and 31%.

The simulation of the sediment during the 2011 and 2012 flood seasons indicates that the closure of the supply canal offtakes in 2012, due to availability of enough rainfall, reduced the deposited sediment by 930 m^3 when compared with 2011. Furthermore, the sediment can be reduced by 29% if the operation of the system will be based on the crop water requirement.

8.4   Sediment management

The water distribution is largely affected by the sediment deposition. However, allocated budgets do not allow implementation of all necessary maintenance works. Options to reduce sediment loads entering and settling in Gezira Scheme include:

- to improve the river basin area through soil conservation measures, construction of sediment detention structures, such as check dams or small reservoirs in the river basin, before the sediment load enters the main river system. This implies that any improvement in the land use practices in the upstream part of the Blue Nile River Basin will reduce the deposition in the scheme. This kind of solution needs joined efforts and can be a long-term solution;
- to install sediment control techniques such as settling basins, exclusion structures, ejection and rejection at the entrance of the main canals or extraction structures within the canal networks. However, these solutions are more effective with coarse sediment and the elimination of sediment is costly.
- a canal design approach to keep sediment in suspension and to pass it to the irrigated fields (Munir, 2011);
- to improve the operation and maintenance for better sediment and water management.

8.5   Conclusions and the way forward

The study found that the setting of and the way in which the control structures are operated have an essential effect on the sedimentation. This effect can be locally or extend several kilometres upstream of the control structures. This has also significant effects on the increase in deposition, since it affects the water depth along the canals. The sediment properties, such as settling velocity, play an essential role in the deposition of fine sediment and there is a high sensitivity of the deposition. The improper desilting process has also essential effects on canal siltation.

Since there is only a limited number of models that deal with cohesive sediment transport in irrigation canals the developed model is a real contribution to the field of cohesive sediment transport. Through the model the possibility of improving the operation and maintenance of the Gezira irrigation canals has been investigated to mitigate the accumulation of sediment along the canals and to minimize the maintenance costs. The model is a useful tool for decision makers and operators to generate a comparative analysis of the sediment transport, addresses the effects of different operation modes on the sedimentation, as well as to attain the most effective sediment and water management. Further development of the FSEDT model by incorporation of consolidation and flocculation processes can be considered.

The night storage system was tested at the minor canal level. It was found to result in more deposition, which requires intensive maintenance works. For Gezira Scheme to adapt with intensification and diversification the best option of operation is to apply the continuous operation system, which can reduce the deposition by 55% when compared to the night storage system.

The analysis shows that most of the deposition occurs between mid-July and the first week of August, which coincides with the period of high sediment concentration. Special consideration has been given to the amount of water delivered during that period. The most effective operation to reduce the deposited amount of sediment and maintenance costs is when the system is operated continuously based on the crop water requirement, especially during the period of high sediment concentration with full capacity of the field outlet pipes. In other words, to adjust the supply to satisfy the crop water demand is the way for better sediment and water management. To attain maximum hydraulic efficiency the canals need to be in an acceptable condition for imperative irrigation management.

Better understanding of the design concept of canals that carry sediment is needed, especially when dealing with cohesive sediment and taking into consideration the criteria for cohesive soils in the design of stable canals. The assumptions in the design need to incorporate the operation of the system. For Gezira Scheme this implies that remodelling will be a more attractive solution than rehabilitation. Furthermore, the controlled water levels in the canals determine the flow through the offtakes, which have to be adjusted to maintain the intended discharge into the canals. In this regard, the cross structures play a critical role in providing the required command, improving the water delivery management and reducing the sediment deposition rate.

Upgrading the monitoring system for the whole system and the development of a computerized database system is needed. Basic data on water levels, gate opening and sediment concentration need to be recorded at representative cross regulators and offtakes. This will help the operators to quickly react in case of water shortage and to easily evaluate the actual situation of the system.

9 References

Aalbers E. E., 2012. Water supply from minor canal to field level in the Gezira irrigation scheme, Sudan. Technical report. Delft University of Technology, Delft, the Netherlands.

Adam H. S., 2005. Agroclimatology, crop water requirment and water management. University of Gezira, Water management and irrigation institute, Khartoum, Sudan.

Ahmed A. A., Adeeb A., El-Monshid B. E. F., 1989. Water mangement: planning for period of water shortage with reference to Gezira Scheme. In: Ahmed A. A. (ed) Irrigation management in the Gezira Scheme, Wad Medani, Sudan., pp. 2-60.

Ahmed A. A., Osman I. S., Babiker A., 2006. Sediment and aquatic weeds management challenges in Gezira Irrigated Scheme.International Sediment Initiative Conference, UNESCO Chair on Water Resource, Friendship Hall, Khartoum, Sudan, pp. 40-46.

Ahmed H. S., Abdelhadi A. W., Takeshi H., 2008. The future of participatory water management in Gezira Scheme, Sudan. International Workshop on Participatory Mangement of Irrigation System, Water Utilization Techniques & Hydrology, 3rd World Water Forum, R(1-10-1), March, Kyoto and Shiga, Japan.

Ali Y. S. A., 2014. The impact of soil erosion in the upper Blue Nile on downstream reservoir sedimentation. PhD thesis. UNESCO-IHE, Delft, the Netherlands.

Ararso G. S., Schultz B., 2008. Planning water management for secure food production in Sub-Saharan Africa. Irrigation and Drainage Published online in Wiley InterScience (www.interscience.wiley.com) DOI: 10.1002/ird.444.

Belaud G., Baume J., 2002. Maintaining equity in surface irrigation network affected by silt deposition. Irrigation and Drainage Engineering 128 (5).

Bhutta M. N., Shahid B. A., Van der Velde E. J., 1996. Using a hydraulic model to prioritize secondary canal maintenance inputs: results from Pujab, Pakistan. Jornal of Irrigation and Drainage Systems 10: 377-392.

Billi P., Ali O. E. B., 2010. Sediment transport of the Blue Nile at Khartoum. Quaternary International Journal, doi:101016/jquaint200911041 .

Bos M. G., 1989. Discharge measurement structures (Third ed.). International Institute for Land Reclamation and Improvement. Wageningen, the Netherlands.

Brush L. M., Ho H. W., Singamsetti S. R., 1962. A study of sediment suspension. Internation Association of Hydrological Sciences, commission Land Erosion (Bari) 59.

Camenen B., Larson M., 2005. A general formula for non-cohesive bed load sediment transport. Estuarine, Coastal and Shelf Science 63: 249-260.

Celik I., Rodi W., 1988. Modeling suspended sediment transport in non-equilibrium situations. Journal of Hydraulic Engineering, 114 (10), 1157-1191.

Central Intelligence Agency (CIA), 2004. World Factbook: Sudan Irrigated Agriculture.The library of Congress Country studies, Cited July 2010 http://wwwphotiuscom/countries/sudan/economy/sudan_economy_irrigated_agricult ur~1784html.

Chan W. Y., Wai O. W. H., Li Y. S., 2006. Critical Shear Stress for Deposition of Cohesive Sediments in Mai Po. Journal of Hydrodynamics, Ser B 18: 300-305 DOI http://dx.doi.org/10.1016/S1001-6058(06)60070-X.

Chow V. T., 1993. Open channel hydraulics Mc Graw Hill International Book Company, Tokio, Japan.

Clemmens A. J., 2006. Improving Irrigated Performance Through an Uderstanding of the Water Delivery Process. Irrigation & Drainage 55: 223-234.

Cole P., Miles G. V., 1983. Two-Dimensional Model of Mud Transport. Hydraulic Engineering 109 (1): 1-12.

Dade W. B., M. N. A. R., Jumars P. A., 1992. Predicting erosion resistance of muds. Marine Geology 105: 255-297.

De Vries M., 1975. A morphological time scale for rivers. Paper presented at the XVIth IAHR Congress. Sao Paulo. Brazil.

Deng Z. Q., Singh V. P., Bengtsson L., 2001. Longitudinal dispersion coefficient in straight rivers. Hydraulic Engineering 127 (11): 919-927.

Depeweg H. W. Th., 1993. Lecture notes on applied hydraulics: gradually varied flow. UNESCO-IHE, Delft, the Netherlands.

Depeweg H. W. Th., Paudel K. P., 2003. Sediment transport problems in Nepal evaluated by SETRIC model. Irrigation and drainage 52: PP 260-274.

Edward T. K., Glysson G. D., 1999. Field methods for measurement of fluvial Sediment. In: Harold P. G., W N. V. (eds) Applications of Hydaulics,Technique of Water- Resources Investigations of the US Geological Survey.

El Monshid B. E. F., El Awad O. M. A., Ahmed S. E., 1997. Environmental effect of the Blue Nile Sediment on the reservoirs and irrigation canals. 5th Nile 2002 Conferance, Addis Ababa.

Elfaki M., 2008. Farmers' participation, innovativeness and initiative the missing ingredient in the world largest farm: Gezira Scheme. Paper presented at the Agriculure Sustainability through Participative Global Extension, Selangor, Malaysia.

Elhassan M., Ahmed E., 2008. Watershed degradation and its impacts on agricultural area and crop production in Gezira Scheme. Paper presented at the Nile Basin Development Forum, Friendship Hall Khartoum, Sudan,17-19 Novembe.r

Ellegaard A. C., Christiansen N., 1994. Laborotary experiments with dredging spoils. Drivation of material-specific parameters for numerical modelling of sediment transport. Report no. 7174. DHI/LIC Joint Venture

Farbrother H. G., 1974. Irrigation practices in Gezira, cotton Research Report, 1970-1971. Cotton Research Corporation. No. 89.

Fischer B. H., List E. j., Koh R. C. Y., Imberger J., Brooks N. H., 1979. Mixing in Inland and Coastal Waters. Academic Press, New York: 104–138.

Galappatti R., 1983. A depth integrated model for suspended transport. Report No. 83-7, Delft University of Technology, Delft, the Netherlands.

Gismalla Y., 2009. Sedimentation Problems in the Blue Nile Reservoirs and Gezira Scheme: A Review. Gezira Journal of Engineering and Applied Sciences 14 (2): 1-12

Gismalla Y. A., Fadul E. M., 2006. Design and remodelling of stable canals in Sudan. International Sediment Initiative Conference, UNESCO Chair on Water Resource, Friendship Hall, Khartoum, Sudan, pp 371-380.

Guan W. B., Wolanski E., Dong L. X., 1998. Cohesive Sediment Transport in the Jiaojiang River Estuary, China. Estuarine, Coastal and Shelf Science 46: 861-871.

Guo, J. C. Y., Hinds J. W., 2013. Numerical criteria for stable kinematic wave overland flow solution. Irrigation and Drainage Engineering JRNIRENG-S-13-00290.

Guy P. H., 1969. Laborotary theory and methods for sediment and analysis. Techneques of Water Resources Investigation of United State Geological Survey, Washington, USA.

Hagos E. Y., 2005. Development and management of irrigation lands in Tigray, Ethiopia. PhD Thesis, Wageningen University and UNESCO-IHE, Delft, the Netherlands, 1-12 pp.

Halcrow Group L., 1998. Guidelines for irrigation canal control, vol 1: user manual, Glasgo, United Kingdom.

Henderson F. M., 1996. Open channel flow Mac Millan Publishing Co. Inc., New York, USA.

HR Wallingford, 1990a. Siltation Monitoring Study.Research for Rehabilitation EX 2244, Hydraulic Research Ltd, Wallingford, Oxfordshire, United Kingdom.

HR Wallingford, 1990b. Sediment Management Study. Research for Rehabilitation EX 2394, Hydraulic Research Ltd, Wallingford, Oxifordshire, United Kingdom, October 1990.

Huang J., Hilldale R. C., Greimann B. P., 2008. Cohesive sediment transport. http://www.usbr.gov/pmts/sediment/kb/ErosionAndSedimentation/chapters/Chapter 4.pdf, Cited July 2010.

Hung M. C., Hsieh T. Y., Wu C. H., Yang J. C., 2009. Two dimentional nonequilibruim non-cohesive and cohesive sediment transport model. Journal of Hydraulic Engineering, 135 (5).

Hussein A. S., Ahmed A. A., Esa S., 1986. Sedimentation problems in the Gezira Scheme. Paper presented at the International conference on Water Resources Needed and Planning in Drought Prone Area, Khartoum, Sudan, December.

Hussin A. S. A., 2006. Sudanese experience in sedimentation engineering. Paper presented at the International Sediment Initiative Conference, Khartoum, Sudan.

Ibrahim A., Stigter C., Adeeb A., Adam H. S., Rheenen W. V., 1999. On-farm sampling density and correction requirements for soil moisture determination in irrigated heavy clay soils in the Gezira, central Sudan. Agricultural Water Management 41: 91-113

Jian H. U., 2008. Study on mathematical modeling for non_uniform sediment transport in anirrigation district along the lower reach of the Yellow River. China Institute of Water Resources and Hydropower Research, Beijing, China .

Jinchi H., Zhaohui W., Eishun Z., 1993. A study on sediment transport in an irrigation discrict. 15th International Congress on Irrigation and Drainage, the Hague, the Netherlands, pp. 1373-1384.

Kashefipour M. S., Falconer R. A., 2002. Longitudinal dispersion coefficients in natural channels. Water Resources Research 36 (6): 1596–1608.

Khan A. Q., 2004. Sustainable use of water resources and irrigation system design. Training Manuals produced by the International Arid Lands Consortium (IALC) and University of Illinois at Urbana Champaign (UIUC), Peshawar, Pakistan.

Kraatz D. B., Mahajan I. K., 1975. Small Hydraulic Structure, FAO Irrigation and Drainage Paper Food and Agriculture Orginization of the United Nation, 26/2, Rome, Italy.

Krone R. B., 1962. Flume studies of the transport of sediment in estuarial shoaling processes. Technical report, Hydraulic Engineering Laboratory, University of California, Berkeley California, USA.

Lawrence P., Atkinson E., Spark P., Counsell C., 2001. Procedure for the selection and outline of canal sediment control structures. Technical Manual, HR Wallingford, United Kingdom.

Lawrence P., Edmund A., 1998. Deposition of fine sediments in irrigation canals. Irrigation and Drainage Systems 12: 371-385.

Ledden M. V., 2003. Sand-mud segregation in estuaries and tidal basin. PhD Thesis, University of Technology Delft, Delft, the Netherlands, pp 23-30.

Liu W.-C., Hsu M.-H., Kuo A. Y., 2002. Modelling of hydrodynamics and cohesive sediment transport in Tanshui River estuarine system, Taiwan. Marine Pollution Bulletin 44: 1076-1088.

Lopes J. F., Dias J. M., Dekeyser I., 2006. Numerical modelling of cohesive sediments transport in the Ria de Aveiro lagoon, Portugal. Journal of Hydrology 319: 176-198

Lui H., 1977. Predicting dispersion coefficient of streams. Environmental Engineering Division 103(1): 59-69.

Lumborg U., 2005. Modelling the deposition, erosion and flux of cohesive sediment through Oresund. Marine Systems 56: 179-193.

Maa J. P.-Y., Kwon J., Hwang K.-N., Ha H.-K., 2008. Critical bed-shear stress for cohesive sediment deposition under steady flows. Technical notes, Journal of Hydraulic Engineering Vol. 134, No. 12.

Mahmoud A., 1999. Proposal for modernisation of the Gezira Irrigation System, Sudan. M.Sc.Thesis UNESCO-IHE, Delft, the Netherlands.

Majumdar H., Carstens M. R., 1967. Diffusion of particles by turbulence: Effect on the particle size. Water Resouces Center, WRC-0967, Georgia Inst. Techn., Atlanta, USA.

Mamad N., 2010. Irrigation performance of Gezira Scheme, Sudan: Assessment of land and water productivity using setellite data. MSc Thesis, UNESCO-IHE , Delft, the Netherlands, 7-8 pp.

Matthews I. S. G., 1952. Investigation of Stable Channels in the Gezira Canalisation Scheme. SID, Tech. Note No.3/52.

Matyukhin V. J., Prokofyev O. N., 1966. Experimental determination of the coefficient of vertical turblent diffusion in water for settling particles. Soviet Hydrol. (Amer. Geophy. Union).Vol.3.

McAnally W. H., Mehta A. J., 2000. Collisional aggregation of fine estuarial sediment. In: William H. M., Ashish J. M. (eds) Proceedings in Marine Science:19-39.

McCave I. N., 2008. Size sorting during transport and deposition of fine sediment: sortable silt and flow speed. In: Camerlenghi M. R. a. A. (ed) Developments in sedimentology Contourites Loon A. J. v., 60.

McCave I. N., Swift S. A., 1976. A physical model for the rate of deposition of fine-grained sediment in the deep sea. Geological Society of America Bulletin 87: 541-546.

Mehta A. J., Letter J., Joseph V., 2013. Comments on the transition between cohesive and cohesionless sediment bed exchange. Estuarine, Coastal and Shelf Science, Vol. 131: 319-324.

Mehta A. J., Partheniades E., 1973. Depositional Behavior of Cohesive Sediments. Technical report No 16, Univ of Florida, Gainesville, Florida, USA.

Mendez N., 1998. Sediment Transport in Irrigation Canals. PhD Thesis, Wageningen Agricultural University and International Institure for Infrastructural, Hydraulic and Environmental Engineering, Delft, the Netherlands.

Milburn D., Krishnappan B. G., 2003. Modelling erosion and deposition of cohesive sediment from Hay River, Northwest Territories, Canada Nordic Hydrology 34 (1-2): 125-138.

Millar R., Quick M., 1998. Stable width and depth of gravel bed Rivers with cohesive banks,. Hydraulic Engineering, 124 (10): 1005-1013.

Munir S., 2011. Role of sediment transport in operation and maintenance of supply and demand based irrigation canals. PhD thesis, UNESCO-IHE, Delft, the Netherlands.

Oklahoma Water Resources Board, 2004. Standard operating procedure for the use of floats to determine stream discharge. Water quality programs division, Oklahoma, USA.

Omer A. Y., 2011. Sedimentation of sand and silt in Roseires Reservoir: vertical and horizontal sorting with time. M.Sc Thesis, UNESCO-IHE Insititute for water education, Delft, the Netherlands.

Parchure T. M., Mehta A. J., 1985. Erosion of soft cohesive sediment deposits. hydraulic Engineering 111(10): 1308-1326.

Partheniades E., 1965. Erosion and deposition of cohesive soils. Journal of the Hydraulic Division Proceedings of the ASCE 91 (HY1): 105-139.

Partheniades E., 1986. The present state of knowledge and needs for future research on cohesive sediment dynamics. In: Wang S. Y., Shen H. W., Ding L. Z. (eds) Third International Symposium on River Sedimentation Pending, University of Mssissippi, USA, pp. 3-25.

Partheniades E., 2009. Cohesive Sediments in Open canals. A Macmillan Publishing Solutions, U.S.A.

Paudel K. P., 2010. Role of sediment in the design and management of irrigation canals. PhD thesis, UNESCO-IHE, Delft, the Netherlands.

Plusquellec H., 1999. The Gezira Irrigation Scheme in Sudan: objective, design and performance. The World Bank Washington, D. C. Paper No. 120, pp. 21-22.

Raudkivi A. J., 1990. Loose Boundary Hydraulics, Third Edition, Pergamon Press, United Kingdom.

Regulation Handbook, 1934. Gezira Canal Regulation Handbook. Ministry of Irrigationnand and Water Resources, Khartoum, Sudan.

Richardson J. F., Zaki W. N., 1954. Sedimentation and fluidization: Part I. Transaction of the onstitution of chemical engineers 32, 35-50.

Rouse H., 1937. Modern conceptions of the mechanics of fluid turbulence. Trans. ASCE, 102, 463-543.

Sahay R. R., 2013. Predicting longitudinal dispersion coefficients in sinuous rivers by genetic algorithm. Hydrology and hydromechanical 61: 214-221.

Sanchez M., Grimigni P., Delane Y., 2004. Steady state vertical distribution of cohesive sediment in a flow. Published by Elsevir SAS, C.R. Geoscience 337 (2005) 357-365.

Sanford L. P., 2008. Modeling a dynamically varying mixed sediment bed with erosion, deposition, bioturbation, consolidation, and armoring. Computers & Geosciences 34: 1263-1283.

Sanford L. P., Halka J. P., 1993. Assessing the paradigm of mutually exclusive erosion and deposition of mud with examples from upper Chesapeake Bay. Marine Geology 114: 37-57.

Scarlatos P. D., Lin L., 1997. Analysis of fine grained sediment movement in small canals. Journal of Hydraulic Engineering Vol. 123, No. 3.

Seo I. W., Cheong T. S., 1998. Predicting longitudinal dispersion coefficient in natural streams. Hydraulic Engineering 124 (1): 25-32.

Setegn S. G., Srinivasan R., Dargahi B., Melesse A. M., 2009. Spatial delineation of soil erosion vulnerability in the LakeTana Basin, Ethiopia. Hydrological Processes Published online in Wiley InterScience (www.interscience.wiley.com) DOI: 10.1002/hyp.7476.

Sherard L., Decker R. S., Ryker N. L., 1972. Piping in earth dams of dispersive clay. Proceedings, American Society of Civil Engineers Specialty Conference on the Performance of Earth and Earth-Supported Structures, vol. 1, pp.589-626.

Sherpa K., 2005. Use of sediment transport model SETRIC in an Irrigation Canal. M.Sc., UNESCO-IHE , Delft, the Netherlands.

Simons D., Senturk F., 1992. Sediment Transport Technology. Water and Sediment Dynamics. Water Resources Publications Colorado, USA.

Smerdon E. T., Beasley R. P., 1959. The Tractive Force Theory Applied to Stability of Open Channels in Cohesive Soils. Research Bulletin 715, University of Missouri, College of Agriculture, Agricultural Experiment Station, Vol. 13, pp. 441–448.

Steenhuis T. S., Collick A. S., Easton Z. M., Leggesse E. S., K. H., Bayabil, White E. D., Awulachew S. B., Adgo E., Ahmed A. A., 2009. Predicting discharge and erosion for the Abay (Blue Nile) with a simple model.

Stowa, MX. Systems, 2002. Duflow Reference Manual. Version 3.5

Sutama N. H., 2010. Mathematical modelling of sediment transport and its improvement in Bekasi Irrigation System, West Java, Indonesia. MSc Thesis, UNESCO-IHE, Delft, the Netherlands.

Sutcliffe J. V., Parks Y. P., 1999. The Hydrology of the Nile IAHS Special Publication no. 5.

Thiruvarudchelvan T., 2010. Irrigation performance of Gezira Scheme in Sudan: Assessment of irrigation efficiency using satellite data. MSc. Thesis, UNESCO-IHE, Delft, the Netherlands.

Thomas W. A., McAnally W. H., 1985. User`manual for the generalized computer program system open-channel flow and sedimentation TABS-2, Department of the Army Waterways Experiment Station. Crops of Engineers, Vicksburg, Mississippi, USA.

Van Rijin L., 1987. Mathematical Modelling Processes in the Case of Suspended Sediment Transport. PhD Thesis, Delft Hydraulic Communication No. 382. Delft, the Netherlands.

Van Rijn L. C., 2007. Unified view of sediment transport by currents and waves. Part II: Suspended transport. Journal of Hydraulic Engineering 133 (6): 668-689.

Van Rijn L. C., Rossum H., Thermes P., 1990. Field verification of 2D and 3D suspended sediment models. journal of Hydraulic Engineering, 116 (10), 1270-1288.

Vinzon S. B., Mehta A. J., 2003. Lutoclines in high concentration estuaries: Some observations at the mouth of the Amazon. Journal of Coastal Research 19 (2): 243-253.

Walworth J., 2006. Soil Structure: The Roles of Sodium and Salts, Department of Soil, Water and Environmental Science, University of Arizona, Arizona, USA.

Wang Z. B., Ribberink J. S., 1986. The validity of a depth-integrated model for suspended sediment transport Journal of Hydraulic Reseach, IAHR, 24(1).

Whitehouse R. J. S., Soulsby R., Roberts W., Mitchener H. J., 2000. Dynamics of Estuarine Muds. Thomas Telford, London, United Kingdom.

William L., Peacockand L., Peter Christensen L., Chapter 15: Interpretation of Soil and Water Analysis, http://iv.ucdavis.edu/files/24409.pdf, pp. 115-120.

Winterwerp J. C., Kesteren W. G. M. V., 2004. Introduction to the physics of cohesive sediment in the marine environment. Elsevier, Delft, the Netherlands.

World Bank, 2000. Options for sustainable development of Gezira Scheme. Report No. 20398-SU, World Bank document, Washington, USA, 21-22 pp.

Wu Y., Falconer R. A., Uncles R. J., 1999. Modelling of Water Flows and Cohesive Sediment Fluxes in the Humber Estuary, UK. Marine Pollution Bulletin 37: 182-189.

Wubneh A. M., 2007. Analysis of sediment transport in the diversion headwork of Shilanat Irrigation Scheme. MSc Thesis, UNESCO-IHE, Delft, the Netherlands.

Yosif D. M., 1993. Determination of settling velocity of cohesive mud in turbulent flow, Ministry of Irrigation and Water Resources-Hydraulic Research Station, April 1993.

Zuwen J., Huib d. V., Chunhong U., 2003. Application of SOBEK model in the Yellow River Estuary. Paper presented at the International Conference on Estuaries and Coasts, Hangzhou, China.

ANNEXES

Annex A.   Symbols

Symbol	Description	Unit
A	Cross-sectional area of flow	m^2
A	Gate opening	m
A_b	Solid area of channel bed	m^2
β	Turbulent Schmidt number	-
B	Bed width	m
C	Sediment concentration	kg/m^3
C_a	Reference sediment concentration	kg/m^3
C_b	Bed sediment concentration	kg/m^3
C_c	Contraction coefficient	-
C_d	Discharge coefficient	-
C_v	Velocity coefficient	-
d or y	Water depth from bed level	m
d_{50}	Median diameter	m
D	Hydraulic diameter	m
D_f	Floc diameter	m
D_i	Deposition flux	kg/m^2/s
E	Erosion rate	kg/m^2/s
e	Degree of accuracy	m
Fr	Froude Number	-
g	Acceleration due to gravity	m/s^2
H	Total head	m
h	Water depth over the crest level	m
i	Subscript for length	M
j	Superscript for time	S
k_r	Von Karman's constant	-
K	Calibration parameter for settling velocity	-
K_a, K_p, K_s	Constants in the regime method	-
K_s	longitudinal dispersion coefficient	m^2/s
m	Side slope	V : 1H
M	Erosion parameter/ re-adaptation parameter	kg/m^2/s
n	Manning's roughness coefficient	s/m$^{1/3}$
P	Wetted perimeter	m
Q	Water discharge	m^3/s
Q_s	Sediment discharge	m^3/s
q	discharge per unit width	m^2/s
R	Hydraulic radius	m
Re_p	Reynolds number of particle	-
S	sediment exchange rate with the bed	kg/m^2/s
S_f	Energy slope	-
S_o	Bed slope	-
t	Time	s
T	Top surface width	m
u	Flow velocity	m/s
u^*	Shear velocity	m/s
u, v, w	Time average velocities in x, y, z coordinates	m/s

Symbol	Description	Unit
w	Crest width	m
w_s	Settling velocity	m/s
W_s	Effective settling velocity	m/s
x, y, z	Distance in rectangular coordinate system	m
y_c	Critical depth	m
y_n	Normal depth	m
z_u	Upstream water level	m
z_w	Crest level	m
z_l	Rouse number	-
Δx	Length Step	m
Δt	Time step	s
V	kinematic viscosity	m^2/s
ϕ	Ratio between concentration and gelling concentration	-
ϕ_s	Volumetric concentration	-
θ	Losses coefficient	-
ε_s	Sediment diffusion coefficient	m^2/s
μ	eddy viscosity	kg/m/s
ρ	Sediment density	kg/m^3
ρ_s	Particle density	kg/m^3
ρ_w	Water density	kg/m^3
σ_r	Courant number	-
τ	Bed shear stress	N/m^2
τ_c	Critical shear stress for erosion	N/m^2
τ_d	Critical shear stress for deposition	N/m^2

Annex B.   Abbreviations and acronyms

ARC	Agricultural Research Corporation
BCM	Billion Cubic Meter
CEC	Cation Exchange Capacity
CIA	Central Intelligence Agency
CWR	Crop Water Requirement
D/S	Downstream
EC	Electrical Conductivity
ESP	Exchangeable Sodium Percentage
ET_0	Reference Evapotranspiration
FSL	Full Supply Level
FSEDT	Fine Sediment Transport
FOP	Field Outlet Pipe
LL	Liquid Limit
MAE	Maximum Absolute Error
ME	Maximum Error
MOIWR	Ministry Of Irrigation and Water Resources
MSL	Mean Sea Level
MW	Movable Weir
NSL	Night Storage Level
NSW/NS	Night Storage Weir
NSS	Night Storage System
PR	Pipe Regulator
PI	Plasticity Index
PL	Plastic Limit
RMSE	Root Mean Square Error
SAR	Sodium Adsorption Ratio
SCC	Sudan Cotton Company
SETRIC	Sediment Transport in Irrigation Canal
SGB	Sudan Gezira Board
SP	Saturation Percentage
U/S	Upstream
WUA	Water Users Association

Annex C. Data collection and flow measurements

Table C.1. Summary of data collection between July and October in 2011 and 2012

Type of measurement	Frequency and locations of collected data
Water level/gate settings	On a daily basis by using a staff gauge. At the measuring locations as in Figure 5.2, in addition to the offtake of Gezira Main Canal.
Diver records	Every 2 hr upstream of the offtake of Zananda and Toman canals in 2011 and 2012. Three more divers were installed D/S the Zananda offtake, Toman offtake and at end of the first reach of Toman Minor Canal.
Flow measurement	Several times at different locations D/S of the head and cross-structures. The measurement varying between 7 and 4 times per location.
Sediment concentration	On daily basis in 2011 and every 2 days in 2012. The frequency reduced at the end of the flood season. The measuring locations as shown in Figure 5.2 in addition to the offtake of Gezira Main Canal.
Grain size distribution for the bed material	At the head reach of Zananda and Toman canals and the tail of the Toman Minor Canal.
Grain size distribution for suspended sediment	At all the measuring locations.
Chemical parameters	Two samples were taken from the head reach of Toman Minor Canal
Bathymetric survey	Four times at the beginning and end of the food season in 2011 and 2012 along Zananda and Toman canals. In addition to several sections before and after desilting to assess the clearance work.
Cropped area	Survey along Toman Minor Canal in 2011 and 2012. Zananda System recorded by SGB
Irrigation schedule (opening of Abu Ishreen canal)	On daily basis (Sept. and Oct. in 2011 and 2012)
Series of discharge and concentration measurements (mass balance)	On daily basis at first reach of Toman Minor Canal
General condition of the system	On daily basis during the study period

C.1 The cropping pattern and irrigation Schedule

The cropped areas in the Zananda Major Canal system were reported by the Sudan Gezira Board in the Tables C.2 and C.3 during the study period. A survey was held to report the exact cropped area per Nimra along Toman Minor Canal as shown in the Tables C.4 and C.5. Table C.6 illustrates the recorded rainfall during July and August. The aim of this data is to detect the rainwater supply.

The command area is divided by Abu Ishreen canals into Nimras (each Nimra has 37.8 ha). It was recognized that the crop intensity is low at some Nimras. However, it is difficult for farmers to follow fixed irrigation intervals and this leads to water mismanagement. The monitoring of the operated FOPs along the first reach and the first three reaches in 2011 and 2012 respectively for a certain period at Toman Minor Canal confirmed that no operation schedule was followed as it is shown in the Tables C.7 and C.8. The Abu Ishreen canals were recorded early in the morning but some might have been opened later during the day.

C.2   Field data used in the model

An intensive data collection program has been conducted to study the deposition process in the irrigation canals and analyse the system by using the modelling approach. In this section the data used for the calibration, validation and simulation of different scenarios are summarized. The inflow and outflow of the major canal under the study are presented in the Tables C.9 and C.10 and applied to the model in time series. The crest levels for the cross structures at 9.1 and 12.5 km from the offtake are also present in Table C.10. The sediment concentrations at the offtakes of Zananda and Toman canals as shown in Table C.11 were set as upstream boundary condition. The canal profiles are presented in the Tables C.12 and C.13.

Table C.2. Total cropped area in 2011 in Zananda Major Canal system

Canals	Cotton (ha)	Sorghum (ha)	Groundnut (ha)	Garden (ha)	Total area (ha)
Gimillia	88	164	50	0	302
G/ Elhosh	144	146	113	46	449
Ballola	15	161	29	23	228
W/ Elmahi	83	186	21	32	322
G/ Abu Gomri	65	46	34	13	157
Toman	65	320	56	20	462
Gemoia Branch	45	37	27	0	109
Gemoia	119	155	155	18	445
W/Husain	131	184	150	4	470
G/Abo Gomri	28	47	23	0	98
Elnifidia Branch	30	56	7	1	94
Elnifidia	26	15	69	3	113
W/Noman U/S K17	118	224	75	15	433
W/ Noman D/S K 9	10	79	32	3	124
Total	966	1820	840	177	3800

(Source: Sudan Gezira Board records)

Table C.3. Total cropped area in 2012 in Zananda Major Canal system

Canals	Cotton (ha)	Sorghum (ha)	Groundnut (ha)	Garden (ha)	Total area (ha)
Zananda	163	3339	655	529	4704
Gimillia		256	139	0	395
G/ Elhosh		241	123	17	381
Ballola		407	92	10	508
W/ Elmahi		512	47	14	575
G/ Abu Gomri		120	29	8	158
Toman		271	95	8	374
Gemoia Branch		281	79	0	360

(Source: Sudan Gezira Board records)

Table C.4. Cropped area per Nimra (ha) in Toman Minor Canal system in 2011

Nimra	Sorghum	Groundnut	Garden	Cotton	Total
1	Fallow				0
2	15	4	5	0	24
3	26	7	0	0	33
4	35	3	1	0	39
5	Fallow				0
6	16	13			29
7	29	3	2	0	34
8	Fallow				0
9	20	3	8	0	31
10				35	35
11	Fallow				0
12	28	8	2	0	37
13	5	0	0	0	5
14	26	7	0	0	33
15	30	4	0	0	34
16	21	1	0	0	22
17	0	0	0	29	29
18	34	2	3	0	38
19	34	2	0	0	36
20	Fallow				0
21	Fallow				0
Total	320	56	20	65	460

Note: Nimra is a command area that irrigated by a field canal (Abu Ishreen)

Table C.5. Cropped area per Nimra (ha) in Toman Minor Canal system in 2012

Nimra	Sorghum	Groundnut	Garden	Cotton	Total
1	35	0	1	0	36
2	Fallow				
3	14	8	12	0	34
4	Fallow				
5	29	1	0	0	29
6	10	1	2	0	13
7	12	2	10	0	24
8	1	3	10	0	14
9	22	0	0	0	22
10	Fallow				
11	27	7	2	0	35
12	Fallow				
13	29	7	0	0	36
14	Fallow				
15	21	8	0	0	29
16	10	5	0	0	15
17	Fallow				
18	Fallow				
19	7	10	3	2	22
20	32	0	0	0	32
21	34	0	0	0	34
Total	283	51	40	2	376

Note: Nimra is a command area that irrigated by field canal (Abu Ishreen)

Table 5.6. Rainfall measurement at K57 from Sennar Dam

2011		2012	
Date	Rainfall (mm)	Date	Rainfall (mm)
2/8	10	4/7	9
5/8	15	10/7	18
9/8	25	13/7	15
14/8	13	22/7	7
18/8	35	25/7	12
20/8	15	29/7	30
29/8	50	1/8	10
		2/8	5
		3/8	13
		4/8	15
		5/8	20
		9/8	30
		31/8	40
Total	163		175

Table C.7. Irrigation schedule in Toman Minor Canal for the first reach

Day/field canal no.	1	2	3	4	5
September					
15		op		op	
16				op	
17			op	op	
18			op	op	
19					
20		op	op	op	
21		op	op	op	
22		op	op	op	
23		op	op	op	
24		op	op	op	
25		op	op	op	
26		op	op	op	
27		op	op	op	
28		op	op	op	
29		op	op	op	
30		op	op	op	
October					
1		op	op	op	
2		op	op	op	
3		op	op		
4		op	op	op	
5		op		op	
6			op		
7		op	op	op	
8				op	
9		op		op	
10		op		op	
11				op	
12		op	op	op	
13		op	op	op	
14				op	
15				op	
16					
17		op			
18		op			
19		op			
20		op			
21		op	op		
22			op		
23			op		
24					
25		op		op	
26		op	op		
27		op		op	
28		op		op	

Note: op: opened FOP, [   ] is fallow

Table C.8. Irrigation Schedule in Toman Minor Canal during September

Day	1st reach					2nd reach					3rd reach					
	1	2	3	4	5	6	7	8	9	10	11	12	13	14	15	16
1			op		op								op		op	op
2			op						op				op		op	op
3			op				op									op
4	op		op				op				op		op			op
5	op										op		op			op
6																op
7																
8								not recorded								
9	op		op		op		op		op		op		op			op
10	op		op		op		op	op	op		op		op			op
11	op		op		op			op	op		op		op			op
12			op					op	op			op	op			op
13			op						op		op		op			op
14									op							op
15								not recorded								
16	op		op					op	op				op			op
17	op		op					op	op							
18	op		op				op									
19	op		op								op					
20	op		op		op				op		op		op			
21								not recorded								
22	op		op		op			op	op		op		op			op
23	op		op					op	op		op		op			
24	op		op						op		op		op			
25	op		op		op			op					op			
26	op		op				op	op			op		op			op
27	op		op				op									
28								not recorded								
29	op		op		op		op		op		op					op
30	op		op		op				op							op

Note: op: opened FOP, [ ] is fallow

Table C.9. Discharge (m³/s) for different canals during the study period 2011

Day	Zananda	G/Elhosh	Gimillia	Ballola	W/Elmahi	Toman	Gemoia	G/Abu Gomri
				July				
1	2.41	0.33	0.17	0.15	0.44	0.26	0.20	0.24
2	3.08	0.36	0.25	0.15	0.44	0.26	0.20	0.24
3	3.46	0.27	0.23	0.15	0.44	0.26	0.24	0.32
4	2.95	0.06	0.06	0.16	0.46	0.32	0.24	0.32
5	3.52	0.21	0.23	0.13	0.47	0.32	0.24	0.32
6	2.41	0.19	0.21	0.14	0.46	0.32	0.24	0.32
7	2.41	0.23	0.25	0.16	0.46	0.35	0.23	0.30
8	2.41	0.28	0.25	0.16	0.46	0.35	0.24	0.32
9	2.88	0.36	0.36	0.16	0.46	0.32	0.26	0.35
10	2.88	0.36	0.31	0.16	0.46	0.32	0.26	0.35
11	2.88	0.24	0.28	0.16	0.46	0.32	0.26	0.35
12	2.64	0.19	0.24	0.16	0.49	0.32	0.26	0.35
13	2.64	0.25	0.33	0.16	0.49	0.32	0.26	0.35
14	2.48	0.24	0.28	0.17	0.49	0.35	0.16	0.17
15	2.93	0.23	0.36	0.17	0.49	0.35	0.16	0.17
16	2.86	0.30	0.25	0.17	0.45	0.35	0.18	0.21
17	2.93	0.27	0.25	0.17	0.45	0.35	0.29	0.41
18	5.43	0.24	0.25	0.15	0.45	0.35	0.29	0.41
19	4.48	0.33	0.30	0.17	0.45	0.35	0.24	0.32
20	5.79	0.33	0.44	0.19	0.53	0.46	0.31	0.46
21	4.48	0.23	0.40	0.19	0.53	0.49	0.31	0.46
22	5.90	0.33	0.54	0.20	0.52	0.46	0.31	0.46
23	5.90	0.33	0.49	0.18	0.51	0.45	0.33	0.50
24	5.90	0.21	0.40	0.17	0.50	0.36	0.29	0.41
25	5.17	0.21	0.44	0.17	0.49	0.32	0.24	0.32
26	6.34	0.08	0.44	0.15	0.50	0.36	0.23	0.30

Sediment and water management of Gezira Scheme, Sudan

Day	Zananda	G/elhosh	Gemilia	Ballola	W/Elmahi	Toman	Gemoia	Abu Gomri
July								
27	6.34	0.23	0.45	0.12	0.41	0.23	0.23	0.30
28	5.17	0.28	0.45	0.16	0.47	0.35	0.23	0.30
29	5.17	0.23	0.40	0.20	0.52	0.39	0.30	0.44
30	5.17	0.21	0.45	0.20	0.52	0.39	0.13	0.13
31	5.17	0.25	0.49	0.20	0.52	0.40	0.29	0.41
August								
1	5.17	0.15	0.34	0.17	0.52	0.60	0.29	0.41
2	6.23	0.34	0.49	0.17	0.52	0.60	0.38	0.61
3	6.23	0.33	0.49	0.17	0.52	0.64	0.38	0.61
4	6.23	0.33	0.49	0.17	0.52	0.60	0.41	0.70
5	6.12	0.21	0.49	0.17	0.52	0.59	0.45	0.79
6	6.12	0.36	0.31	0.17	0.38	0.59	0.41	0.70
7	6.12	0.37	0.31	0.17	0.38	0.69	0.44	0.77
September								
10	1.27			0.15	0.49	0.09	0.00	0.00
11	1.27			0.13	0.44	0.09	0.07	0.05
12	1.27			0.10	0.43	0.06	0.00	0.00
13	1.34			0.12		0.24	0.14	0.14
14	1.63					0.39	0.18	0.21
15	2.13			0.13	0.41	0.39	0.18	0.21
16	2.64			0.13	0.52	0.39	0.16	0.17
17	2.64	0.13	0.19	0.17	0.50	0.39	0.26	0.35
18	2.64			0.20	0.50	0.54	0.29	0.41
19	2.64	0.15	0.19	0.20	0.50	0.62	0.29	0.41
20	2.64			0.19	0.50	0.57	0.29	0.41
21	2.64			0.20	0.50	0.59	0.29	0.41
22	2.64	0.11	0.23	0.23	0.53	0.54	0.38	0.61

Day	Zananda	G/elhosh	Gemilia	Ballola	W/ Elmahi	Toman	Gemoia	Abu Gomri
				September				
23	2.64	0.11	0.23	0.23	0.53	0.54	0.40	0.65
24	4.64	0.08	0.25	0.23	0.53	0.54	0.40	0.67
25	4.64	0.10	0.33	0.21	0.52	0.54	0.38	0.61
26	4.64	0.08	0.23	0.17	0.49	0.36	0.29	0.41
27	4.64	0.06	0.17	0.15	0.47	0.35	0.22	0.27
28	4.64	0.12	0.28	0.23	0.52	0.42	0.30	0.44
29	4.64	0.17	0.33	0.23	0.53	0.52	0.32	0.48
30	4.64	0.17	0.33	0.23	0.53	0.43	0.29	0.41
				October				
1	4.64	0.17	0.33	0.23	0.53	0.45	0.16	0.17
2	4.64	0.19	0.36	0.23	0.53	0.46	0.16	0.17
3	4.64	0.20	0.33	0.24	0.53	0.65	0.16	0.17
4	4.64	0.25	0.40	0.26	0.53	0.08	0.13	0.13
5	4.64	0.19	0.33	0.24	0.54	0.39	0.11	0.10
6	4.64	0.21	0.37	0.27	0.53	0.35	0.11	0.10
7	4.64	0.21	0.34	0.27	0.53	0.35	0.11	0.10
8	4.64	0.21	0.33	0.23	0.50	0.26	0.11	0.10
9	4.64	0.21	0.33	0.23	0.50	0.26	0.11	0.10
10	4.64	0.13	0.25	0.18	0.50	0.31	0.11	0.10
11	4.64	0.11	0.33	0.13	0.49	0.40	0.17	0.19
12	4.64	0.23	0.40	0.17	0.51	0.48	0.20	0.24
13	4.64	0.14	0.31	0.18	0.52	0.23	0.13	0.13
14	4.64	0.24	0.4	0.23	0.53	0.20	0.13	0.13
15	4.64	0.33	0.49	0.23	0.54	0.20	0.16	0.17
16	4.64	0.36	0.52	0.23	0.54	0.20	0.16	0.17
17	4.64	0.30	0.54	0.23	0.56	0.20	0.03	0.02
18	4.64	0.11	0.19	0.23	0.56	0.26	0.00	0.00

Sediment and water management of Gezira Scheme, Sudan

Day	Zananda	G/elhosh	Gemilia	Ballola	W/Elmahi	Toman	Gemoia	Abu Gomri
				October				
19	4.64	0.11	0.19	0.18	0.50	0.26	0.00	0.00
20	4.64	0.10	0.19	0.18	0.50	0.26	0.00	0.00
21	4.64	0.13	0.18	0.18	0.50	0.26	0.00	0.00
22	4.64	0.17	0.18	0.18	0.50	0.26	0.00	0.00
23	4.64	0.21	0.15	0.18	0.50	0.26	0.00	0.00
24	4.64	0.19	0.15	0.18	0.50	0.26	0.00	0.00
25	4.64	0.21	0.18	0.18	0.50	0.26	0.07	0.05
26	4.64	0.18	0.26	0.18	0.50	0.26	0.07	0.05
27	4.64	0.14	0.33	0.09	0.38	0.26	0.00	0.00
28	4.64	0.14	0.24	0.00	0.27	0.26	0.00	0.00
29	2.41	0.00	0.19	0.00	0.27	0.26	0.00	0.00

Table C.10. Discharge (m³/s) for different canals and crest level (CL) (m+MSL) for the control structures at 9.1 and 12.5 km from the offtake during the study period 2012

July

Day	Zananda	Gimillia	G/ Elhosh	Bellola	W/ Elmahi	A/Gomri	Gemoia	Toman	CL 9.1	CL 12.5
1	3.51						0.42	0.26		
2	3.51						0.43	0.29		
3	3.51	0.43	0.22	0.33	0.39	0.15	0.49	0.68	411.55	410.43
4	3.51	0.40	0.38	0.37	0.49	0.05	0.46	0.51	411.73	410.43
5	3.51	0.40	0.37	0.38	0.53	0.03	0.46	0.47	411.73	410.43
6	3.51	0.37	0.30	0.36	0.43	0.01	0.45	0.42	411.73	410.43
7	3.51	0.33	0.22	0.33	0.34	0.23	0.49	0.42	411.73	410.43
8	3.51	0.34	0.24	0.33	0.43	0.26	0.58	0.20	411.73	410.43
9	4.36	0.16	0.47	0.40	0.50	0.06	0.61	0.28	411.73	410.43
10	4.36	0.17	0.37	0.35	0.50	0.00	0.52	0.34	411.65	410.43
11	4.36	0.16	0.25	0.31	0.43	0.00	0.56	0.10	411.55	410.43
12	4.36	0.15	0.32	0.22	0.64	0.00	0.38	0.09	411.55	410.43
13	4.19	0.14	0.30	0.13	0.35	0.00	0.20	0.07	411.55	410.43
14	4.19	0.12	0.27	0.13	0.05	0.00	0.20	0.07	411.55	410.43
15	5.17	0.28	0.40	0.15	0.05	0.02	0.20	0.07	411.55	410.43
16	4.03	0.28	0.40	0.14	0.06	0.05	0.20	0.07	411.55	410.43
17	3.88	0.28	0.40	0.22	0.06	0.05	0.24	0.10	411.55	410.43
18	3.88	0.17	0.33	0.17	0.05	0.05	0.20	0.07	411.55	410.43
19	4.07	0.17	0.30	0.17	0.05	0.02	0.16	0.07	411.58	410.43
20	4.07	0.15	0.29	0.17	0.05	0.02	0.16	0.09	411.58	410.43
21	3.35	0.12	0.28	0.17	0.05	0.02	0.16	0.10	411.58	410.43
22	2.31	0.12	0.30	0.19	0.05	0.03	0.20	0.12	411.58	410.43
23	2.10	0.04	0.21	0.13	0.00	0.00	0.05	0.02	411.58	410.43
24	2.12	0.00	0.16	0.00	0.00	0.00	0.07	0.04	411.58	410.43
25	2.66	0.00	0.06	0.00	0.00	0.00	0.00	0.02	411.58	410.43

Day	Zananda	Gimillia	G/ Elhosh	Bellola	W/ Elmahi	A/Gomri	Gemoia	Toman	CL 9.1	CL 12.5
July										
26	2.66	0.00	0.00	0.00	0.00	0.00	0.00	0.01	411.58	410.43
27	1.07	0.00	0.00	0.00	0.00	0.00	0.00	0.00	411.58	410.43
28	1.07	0.00	0.00	0.00	0.00	0.00	0.00	0.00	411.58	410.43
29	1.04	0.00	0.00	0.00	0.00	0.00	0.00	0.00	411.58	410.43
30	0.77	0.00	0.00	0.00	0.00	0.00	0.00	0.00	411.58	410.43
31	0.77	0.00	0.00	0.00	0.00	0.00	0.00	0.00	411.58	410.90
August										
1	0.77	0.00	0.00	0.00	0.00	0.00	0.00	0.00	411.75	410.80
2	0.77	0.00	0.00	0.00	0.00	0.00	0.00	0.00	411.75	410.80
3	0.77	0.00	0.00	0.00	0.00	0.00	0.00	0.00	411.75	410.85
4	0.77	0.00	0.00	0.00	0.00	0.00	0.00	0.00	411.75	410.85
5	0.77	0.00	0.00	0.00	0.00	0.00	0.00	0.00	411.65	410.85
6	0.77	0.00	0.00	0.00	0.00	0.00	0.00	0.00	411.65	410.90
7	0.50	0.00	0.00	0.00	0.00	0.00	0.00	0.00	411.65	410.88
8	0.50	0.00	0.00	0.00	0.00	0.00	0.00	0.00	411.65	410.80
9	0.50	0.00	0.00	0.00	0.00	0.00	0.00	0.00	411.65	410.88
10	1.07	0.00	0.00	0.00	0.00	0.00	0.00	0.00	411.65	410.75
11	1.07	0.00	0.00	0.00	0.00	0.00	0.00	0.00	411.75	410.57
24	0.97	0.00	0.00	0.00	0.00	0.00	0.00	0.00	411.70	410.85
25	2.52	0.08	0.07	0.08	0.05	0.00	0.07	0.01	411.70	410.85
26	2.52	0.28	0.08	0.09	0.28	0.00	0.20	0.14	411.70	410.85
27	2.52	0.28	0.08	0.09	0.28	0.00	0.20	0.12	411.70	410.85
28	2.13	0.04	0.06	0.05	0.15	0.00	0.16	0.04	411.70	410.65
29	2.13	0.14	0.16	0.31	0.28	0.17	0.38	0.08	411.67	410.65
30	2.13	0.00	0.00	0.09	0.00	0.00	0.61	0.03	411.67	410.65
31	1.78	0.11	0.06	0.12	0.11	0.06	0.43	0.11	411.70	410.65

Day	Zananda	Gimillia	G/ Elhosh	Bellola	W/ Elmahi	A/Gomri	Gemoia	Toman	CL 9.1	CL 12.5
					September					
1	2.96	0.22	0.11	0.15	0.21	0.11	0.24	0.20	411.70	410.65
2	5.94	0.37	0.13	0.15	0.21	0.15	0.24	0.23	411.70	410.65
3	4.10	0.28	0.13	0.33	0.21	0.15	0.24	0.30	411.52	410.40
4	7.20	0.48	0.18	0.24	0.28	0.06	0.61	0.28	411.70	410.40
5	3.85	0.34	0.16	0.24	0.15	0.00	0.65	0.10	411.70	410.40
6	3.85	0.34	0.27	0.14	0.28	0.03	0.66	0.18	411.70	410.40
7	3.85	0.28	0.28	0.19	0.15	0.04	0.57	0.25	411.70	410.40
8	2.94	0.21	0.29	0.24	0.02	0.06	0.48	0.31	411.70	410.40
9	2.94	0.21	0.27	0.24	0.26	0.24	0.45	0.40	411.70	410.40
10	2.94	0.21	0.25	0.22	0.25	0.06	0.56	0.54	411.70	410.40
11	3.57	0.33	0.30	0.40	0.52	0.06	0.58	0.59	411.70	410.40
12	3.57	0.29	0.19	0.31	0.52	0.06	0.56	0.35	411.70	410.40
13	3.57	0.30	0.19	0.31	0.52	0.06	0.47	0.36	411.70	410.40
14	3.57	0.21	0.16	0.30	0.52	0.25	0.47	0.41	411.70	410.40
15	3.57	0.29	0.13	0.30	0.52	0.44	0.47	0.47	411.70	410.40
16	3.57	0.38	0.15	0.39	0.52	0.44	0.59	0.47	411.70	410.40
17	3.57	0.49	0.19	0.39	0.52	0.44	0.58	0.54	411.70	410.40
18	3.57	0.35	0.13	0.31	0.64	0.35	0.56	0.63	411.70	410.40
19	3.85	0.32	0.12	0.28	0.53	0.35	0.54	0.63	411.70	410.40
20	3.85	0.30	0.12	0.27	0.53	0.41	0.56	0.76	411.70	410.40
21	3.71	0.32	0.16	0.31	0.53	0.37	0.56	0.65	411.70	410.40
22	3.57	0.33	0.21	0.35	0.52	0.32	0.56	0.55	411.70	410.40
23	3.57	0.33	0.21	0.35	0.52	0.40	0.61	0.39	411.70	410.75
24	3.57	0.33	0.21	0.35	0.52	0.40	0.61	0.39	411.70	410.75
25	3.57	0.33	0.21	0.35	0.52	0.26	0.47	0.58	411.70	410.75
26	3.57	0.29	0.17	0.35	0.52	0.44	0.45	0.42	411.70	410.45
27	3.57	0.27	0.17	0.33	0.52	0.00	0.42	0.42	411.70	410.52

Day	Zananda	Gimillia	G/Elhosh	Bellola	W/Elmahi	A/Gomri	Gemoia	Toman	CL 9.1	CL 12.5
					September					
	3.57	0.38	0.19	0.35	0.52	0.00	0.45	0.44	411.70	410.52
29	3.57	0.49	0.22	0.38	0.52	0.00	0.47	0.47	411.70	410.52
30	3.57	0.46	0.22	0.38	0.52	0.00	0.47	0.47	411.70	410.52
					October					
1	3.57	0.46	0.22	0.38	0.52	0.00	0.56	0.50	411.70	410.52
2	4.10	0.28	0.18	0.33	0.52	0.00	0.42	0.29	411.70	410.52
3	4.10	0.36	0.21	0.38	0.52	0.00	0.56	0.58	411.70	410.52
4	4.10	0.87	0.20	0.50	0.52	0.00	0.61	0.58	411.70	410.40
5	4.10	0.88	0.22	0.53	0.52	0.00	0.63	0.58	411.70	410.40
6	4.10	0.90	0.23	0.56	0.52	0.00	0.66	0.58	411.70	410.40
7	3.85	0.93	0.28	0.59	0.52	0.00	0.71	0.58	411.70	410.40
8	3.85	0.84	0.25	0.50	0.52	0.00	0.71	0.58	411.70	410.40
9	3.85	0.79	0.23	0.13	0.52	0.00	0.71	0.61	411.70	410.40
10	3.85	0.12	0.37	0.13	0.00	0.00	0.66	0.58	411.35	410.40
11	5.40	0.12	0.37	0.13	0.00	0.00	0.66	0.35	411.35	410.40
12	2.97	0.12	0.37	0.13	0.00	0.00	0.66	0.35	411.35	410.40
13	2.97	0.12	0.37	0.13	0.00	0.00	0.66	0.35	411.35	410.40
14	2.64	0.12	0.37	0.00	0.00	0.00	0.69	0.43	411.35	410.40
15	2.64	0.00	0.00	0.00	0.00	0.00	0.69	0.43	411.68	410.40
16	3.45	0.19	0.00	0.00	0.34	0.00	0.63	0.43	411.68	410.40

Table C.11. Sediment concentration (ppm) at the offtakes of Zananda and Toman canals

Day	Zananda	Toman		Day	Zananda	Toman
July 2011				July 2012		
1	2690	1940		1	788	2600
2	2250	1460		2	815	4330
3	2080	1740		3	815	3370
4	1930	2760		4	487	5110
5	1630	4660		5	487	5110
6	2030	2350		6	3700	4330
7	2120	1430		7	5000	4330
8	1860	2730		8	5100	7300
9	1660	828		9	2990	7300
10	1400	2020		10	4460	6410
11	1550	2750		11	4550	6410
12	1840	3480		12	5610	7250
13	4150	3870		13	8670	8090
14	4050	4730		14	8480	8090
15	4280	4070		15	8480	9960
16	4230	5400		16	9270	9960
17	9640	11200		17	9270	8190
18	7640	8380		18	9770	8190
19	11000	11600		19	9770	8960
20	11100	9420		20	9850	9740
21	11300	10400		21	9850	9740
22	7410	11500		22	9720	8760
23	7270	11500		23	9720	8760
24	7230	8493		24	8230	7670
25	13800	10500		25	8230	7670
26	14000	11300		26	13900	closed
27	9470	8710		27	13900	Closed
28	8550	9410		28	14400	Closed
29	6160	7400		29	14400	Closed
30	6000	6990		30	8200	Closed
31	3630	6160		31	8200	Closed

Date	Zananda	Toman
August 2011		
1	2910	3080
2	2960	closed
3	7370	closed
4	6800	closed
5	10200	closed
6	11200	closed
7	10500	closed
September 2011		
5	3970	closed
6	3970	closed
7	3970	closed
8	3200	closed
9	3200	closed
10	2930	4845
11	2930	4845
12	2810	2116
13	2810	2116
14	2810	1863
15	1200	1540
16	1200	1540
17	1130	1540
18	1300	1540
19	1400	1708
20	1400	1708
21	1400	1708
22	1400	1294
23	1400	1294
24	1680	1294
25	1680	1294
26	1180	1289
27	1180	1289
28	1180	1289
29	993	901
30	993	901

Date	Zananda	Toman
August 2012		
1	5640	Closed
2	5640	Closed
3	3900	Closed
4	3900	Closed
5	3260	Closed
6	3260	Closed
7	3260	Closed
8	3890	Closed
9	4200	2265
10	4520	Closed
11	4520	1880
24	3510	620
25	3510	5800
26	4300	5800
27	4300	3460
28	4360	3460
29	4360	4180
30	3390	4020
31	3390	3860
September 2012		
1	3550	3860
2	3550	3540
3	5180	3540
4	5180	2840
5	3300	2840
6	3300	2380
7	3270	1930
8	1750	1930
9	1330	1930
10	1330	1930
11	1460	1020
12	1460	1020
13	1290	1020
14	1290	1020

Date	Zananda	Toman
October 2011		
1	1000	791
2	1000	791
3	953	791
4	953	791
5	953	791
6	619	421
7	619	421
8	409	421
9	409	421
10	324	347
11	324	347
12	324	347
13	400	351
14	400	351
15	360	351
16	360	351
17	360	351
18	360	323
19	360	323
20	360	323
21	360	323
22	326	303
23	326	303
24	326	303
25	326	303
26	326	258
27	326	258
28	326	258
29	268	227
30	268	227
31	268	227

Date	Zananda	Toman
September 2012		
15	2580	590
16	2580	590
17	2580	590
18	2580	590
19	2580	590
20	2580	376
21	2280	161
22	2280	161
23	419	161
24	419	437
25	684	437
26	684	437
27	874	437
28	874	437
29	483	437
30	483	136
October 2012		
1	678	136
2	678	349
3	701	349
4	701	283
5	783	217
6	783	217
7	694	286
8	694	286
9	308	110
10	308	110
11	307	199
12	251	287
13	305	287
14	196	287
15	162	287

Table C.12. Bed profile of Zananda Major Canal before and after the flood seasons in 2011 and 2012

2011			2012			
Abscissa (m)	Bed level in June (m+MSL)	Bed level in November (m+MSL)	Abscissa (m)	Bed level in June (m+MSL)	Abscissa (m)	Bed level in October (m+MSL)
200	412.38	412.47	200	412.46	200	412.71
400	412.41	412.66	400	412.52	400	412.69
600	412.25	412.49	600	412.41	600	412.59
800	412.37	412.59	800	412.44	800	412.61
1000	412.35	412.48	1000	412.38	1000	412.50
1200	412.17	412.43	1200	412.38	1200	412.47
1400	412.25	412.42	1400	412.37	1400	412.43
1600	412.27	412.38	1600	412.38	1600	412.45
1800	412.34	412.24	1800	412.41	1800	412.43
2000	412.17	412.51	2000	412.36	2000	412.41
2200	412.28	412.38	2200	412.32	2200	412.41
2400	412.14	412.26	2400	412.24	2400	412.40
2600	412.14	412.39	2600	412.21	2600	412.39
2800	412.20	412.39	2800	412.31	2800	412.40
3000	412.08	412.16	3000	412.18	3000	412.32
3200	412.10	412.24	3200	412.24	3200	412.37
3400	412.01	412.17	3400	412.13	3400	412.23
3600	412.02	412.17	3600	412.17	3600	412.28
3800	411.89	411.86	3800	412.07	3800	412.21
4000	411.92	412.11	4000	412.05	4000	412.21
4200	411.88	412.06	4200	412.13	4200	412.21
4400	411.87	412.16	4400	412.16	4400	412.29
4600	411.89	412.07	4600	412.10	4600	412.22
4800	411.84	412.06	4800	412.12	4800	412.21
5000	411.79	411.98	5000	411.97	5000	412.11
5200	411.76	411.96	5200	411.94	5200	412.10
5400	411.79	411.97	5400	412.03	5400	412.14
5600	411.77	411.92	5600	411.96	5600	412.09
5800	411.72	411.99	5800	411.91	5800	412.02
6000	411.60	411.86	6000	411.83	6000	412.01
6200	411.64	411.84	6200	411.80	6200	412.01
6400	411.56	411.76	6400	411.83	6400	411.93
6600	411.56	411.79	6600	411.83	6600	411.92
6800	411.50	411.67	6800	411.78	6800	411.82
7000	411.50	411.59	7000	411.72	7000	411.81
7200	411.44	411.63	7200	411.71	7200	411.80
7400	411.56	411.52	7400	411.76	7400	411.76
7600	411.40	411.42	7600	411.61	7600	411.71
7800	411.40	411.51	7800	411.58	7800	411.68

2011			2012			
Abscissa (m)	Bed level in June (m+MSL)	Bed level in November (m+MSL)	Abscissa (m)	Bed level in June (m+MSL)	Abscissa (m)	Bed level in October (m+MSL)
8000	411.32	411.47	8000	411.60	8000	411.66
8200	411.34	411.47	8200	411.51	8200	411.60
8400	411.37	411.47	8400	411.49	8400	411.63
8600	411.39	411.48	8600	411.52	8600	411.61
8800	411.31	411.34	8800	411.40	8800	411.46
9000	411.31	411.30	9000	411.46	9000	411.47
9135	411.20	411.50	9200	411.12	9200	411.12
9335	411.19		9600	410.89	9600	410.97
9935	411.04		10000	410.63	10000	410.72
10000	410.89		10400	410.57	11000	410.60
10400	410.84		10800	410.56	11400	410.62
10800	410.66		11200	410.48	11800	410.50
11200	410.67		11600	410.35	12000	410.50
11600	410.54		12000	410.45	12400	410.63
12000	410.60		12400	410.56		
12170	410.72					

Table C.13. The cross-section survey of Toman Minor Canal

| 2011 | | 2012 | | |
Abscissa (m)	Bed level in June (m+MSL)	Abscissa (m)	Bed level in June (m+MSL)	Bed level in October (m+MSL)
0	409.65	0	410.52	410.73
200	409.61	200	410.59	410.73
400	409.57	400	410.60	410.72
600	409.53	600	410.38	410.72
800	409.49	800	410.46	410.65
1000	409.45	1000	410.35	410.54
1200	409.41	1200	410.20	410.34
1200	409.37	1400	410.16	410.32
1400	409.33	1600	410.06	410.21
1600	409.29	1800	410.03	410.21
1800	409.25	2000	410.00	410.16
2000	409.21	2200	409.93	410.10
2200	409.17	2400	409.95	410.09
2400	409.13	2600	409.93	410.08
2600	409.09	2800	409.86	409.99
2800	409.05	3000	409.81	409.98
3000	409.01	3200	409.82	409.96
3000	409.01	3400	409.81	409.95
3200	408.99	3600	409.74	409.87
3400	408.97	3800	409.62	409.81
3600	408.95	4000	409.56	409.74
3800	408.93	4200	409.57	409.77
4000	408.91	4400	409.61	409.72
4200	408.89	4600	409.62	409.73
4400	408.87	4800	409.31	409.47
4600	408.85	5000	409.38	409.52
4700	408.84	5200	409.25	409.39
4700	408.84	5400	409.20	409.34
4800	408.81	5600	409.22	409.35
5000	408.76	5800	409.10	409.23
5200	408.71	6000	409.17	409.30
5400	408.66			
5600	408.61			
5800	408.56			
6000	408.51			

Annex D. Calibration of hydraulic structures

There are numerous physical factors that affect the overall performance of the irrigation system. One of these factors is the deviation of the structure calibration from its original form, thus changes the hydraulic properties of the structure itself. This may be due to the change in bed level, cross-section, bed slope and bed roughness. The calibration of the hydraulic structures has been derived from the direct measurements of the canal discharges and accurate data on the water levels and gate settings. Calibration of measuring structures is required to establish a relationship between water stage and discharge for any water depth or gate openings in case of sluice gates. Measurements are done in a rating section of known dimensions.

D.1 Sluice gate structure

The flow through the offtake of Zananda Major Canal is governed by the orifice equation. The submerged flow is governed by the following equation:

$$Q = C_d BO \sqrt{2g\Delta h} \tag{D.1}$$

Or can be written as:

$$Q = C\,B\,O\sqrt{\Delta h} \tag{D.2}$$

Where:
Q   = discharge (m^3/s)
C_d = discharge coefficient
B   = width of the gate (m)
O   = opening of the gates (m)
Δh = different in head between upstream and downstream (m)
g   = gravity acceleration (m/s^2)
C   = overall coefficient

The discharge coefficient C_d is determined by actual measurement directly in the field, which implicitly incorporates all the effects (very sensitive to low flow head values). The flow measurements have been conducted at different flow conditions, as flow conditions may vary widely depending on gate openings and differential head (Bos, 1989).

In Gezira Scheme the field engineer at the group regulators at 57 km from Sennar Dam applies Equation (D.2) with C = 3 to compute the authorize discharges of Zananda Major Canal. The relationship between computed discharges (Q_c) and measured discharges (Q_m) during the flood season of 2011 and 2012 gives a high correlation coefficient 0.95 as shown in Figure D.1. Based on Figure D.2 the developed equation is:

$$Q = (3.69O + 0.5)\sqrt{\Delta h} \tag{D.3}$$

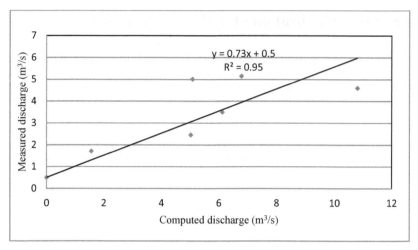

Figure D.1. Relationship between computed discharge and measured discharge at Zananda offtake structure

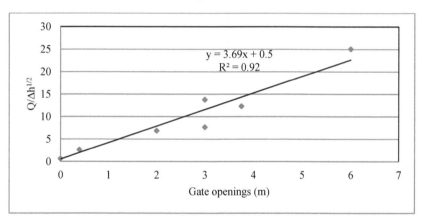

Figure D.2. Correlation between discharge (Q), gate opening and head difference (Δh) at Zananda offtake structure

D.2   Movable weir offtake structures

D.2.1   The standard movable weir Series-II (MW-II)

The standeard movable weir Series-II (MW-II) is used for discharge between 1 m³/s and 5 m³/s and maximum travelling distance of 0.8 m as presented in Table 3.6. The governing equation is as follows:

$$Q = 2.3 \, w \, h^{1.6}$$

(D.4)

Where:
h   = water depth above crest level (m)
w   = is the weir width (m)

The first cross structure in Zananda Major Canal is a movable weir at about 9.1 km from the offtake and has a travel distance of about 0.9 m, while the second one at 12.5 km from the offtake has a maximum travel distance of 1.0 m, different from the standard. The flow measurement for the first cross structure shows that the weir is in an acceptable condition and Equation (D.4) can be applied to estimate the discharge.

A stage discharge relationship has been developed for the second cross structure and the following equation has been obtained as shown in Figure D.3:

$$Q = 2.17 w h^{1.35} \tag{D.5}$$

A low correlation coefficient is observed betweenthe measured discharge and the water depth above the crest level at W/ Elmahi Minor Canal offtake as shown in Figure D.4. This may be referred to the limited data. Therefore Equation (D.4) was applied in this case.

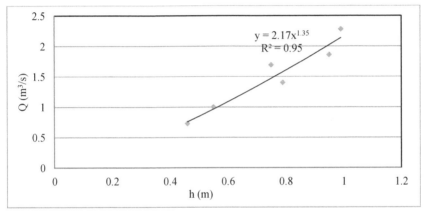

Figure D.3. Stage discharge relationship at the second cross structure (K 12.5) in Zananda Major Canal

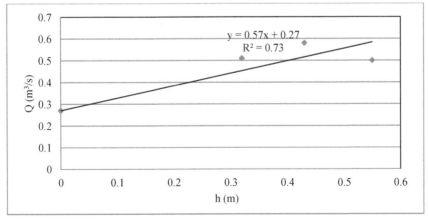

Figure D.4. Relation between measured flow (Q) and water depth above the crest level (h) at W/Elmahi Minor Canal offtake

D.2.2   The standard movable weir Series-I (MW-I)

The standard movable weir Series-I (MW-I) type of structures have been installed in most of the offtakes of the minor canals in Gezira Scheme. They are used for discharge up to 1 m³/s and a maximum travel distance of 0.6 m. The governing equation of this type is:

$$Q = 2.18 \, w \, h^{1.6}$$

(D.6)

A relation between discharge and head above the crest level has been developed for Ballola, Gimillia and G/Abu Gomri minor canals, since the hydraulic characteristics of the canals are similar. The crest width is 0.8 m for the three offtake structures. The flow equation has been developed from the relation in Figure D.5:

$$Q = 0.7 w h^{1.6} + 0.05$$

(D.7)

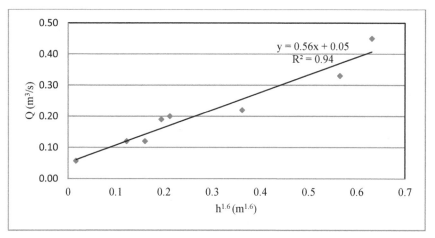

Figure D.5. Relation between measured discharge and water depth over the crest level for Ballola, Gimillia and G/Elhosh minor canals

In case of overflow above the fixed beam (Figure 5.17), the fixed beam is treated as a sharp crested weir and the flow over the weir is based on Equation (D.8) (Paudel, 2010). The total flow is the sum of Equations (D.7) and (D.8):

$$Q = \frac{2}{3} C_d w \sqrt{2g} \, h^{1.5}$$

(D.8)

Where C_d computed based on empirical relation as follows (Paudel, 2010):

$$C_d = 0.611 + 0.08 \frac{h}{p}$$

(D.9)

Where: P is height of crest.

The frame of the MW-I at the offtake of Gemoia Minor Canal has been broken and farmers use sacks to control the water demand as shown in Figure D.6. The discharge and the head above the crest is measured and the following relation has been obtained from Figure D.7.

$$Q = 0.9 \, w \, h^{1.09} \tag{D.10}$$

Figure D.6. Gemoia offtake structure at low irrigation demand closes by sacks

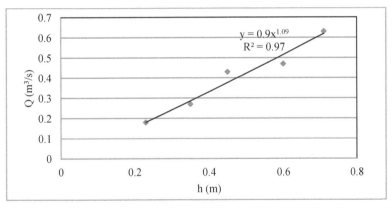

Figure D.7. The relation between water depth over crest level (h) and measured discharge at the offtake of Gemoia Minor Canal

Furthermore, two flow conditions have been recognized in Toman Minor Canal (crest width 0.8 m). The normal situation when the flow is below the fixed beam and the other, when the water depth exceeds the top of the fixed beam (overflow). The developed equations based on the two cases are shown in the Figures D.8 and D.9. Table D.1 summarize the measuring data for different control structures.

- In case of overflow above the fixed beam:

$$Q = 1.79 \, w \, h^{1.74} \tag{D.11}$$

- The water depth not exceed the level of the fixed beam:

$$Q = 1.51h^{1.6} + 0.045 \qquad\qquad\qquad (D.12)$$

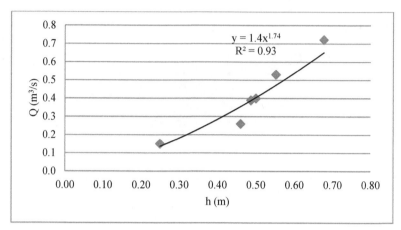

Figure D.8. Measured discharge and water depth above the crest in case of flow over the fixed beam at Toman Minor Canal

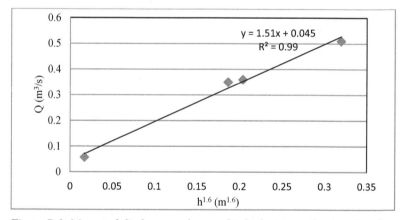

Figure D.9. Measured discharge and water depth above crest in case of no flow over the fixed beam at Toman Minor Canal

Modular flow was recognized during most of the flow measurements for the overflow control structures. The submergence was checked by measuring the difference between upstream or downstream water level and crest level, then h_2/h_1 was checked. These relations were developed based on the collected data. For more reliable discharge relations more measurements will be needed under different flow conditions, taking into account the accuracy of the measurements.

Table D.1. Results of the flow measurement for the different canals

Canal offtakes	Date	Q measured (m^3/s)	Opening/depth over MW (m)
Zananda	18/07/2011	4.60	3.0
	13/09/2011	1.71	0.40
	28/09/2011	5.01	6.00
	10/07/2012	5.15	3.00
	12/08/2012	0.50	0.00
	06/09/2012	2.70	3.75
	11/10/2012	4.35	3.75
	16/10/2012	3.50	2.00
G/Elhosh, Gimillia and Ballola	07/07/2011	0.19	0.36
	18/07/2011	0.20	0.38
	13/09/2011	0.12	0.27
	28/09/2011	0.22	0.53
	10/07/2012	0.45	0.75
	10/07/2012	0.33	0.70
	29/9/2012	0.12	0.32
W/ Elmahi	18/07/2011	0.51	0.32
	13/09/2011	0.27	0.00
	28/09/2011	0.58	0.43
	10/07/2012	0.50	0.55
Toman	07/07/2011	0.36	0.37
	13/09/2011	0.26	0.25
	28/09/2011	0.51	0.49
	13/09/2012	0.35	0.35
Toman- case of overflow	12/9/2011	0.15	0.25
	16/09/2011	0.26	0.46
	19/9/2011	0.39	0.49
	20/9/2011	0.72	0.68
	5/10/2011	0.53	0.55
	27/09/2012	0.40	0.50
Gemoia	07/07/2011	0.27	0.35
	13/09/2011	0.18	0.23
	28/09/2011	0.43	0.45
	13/09/2012	0.47	0.60
	16/10/2012	0.63	0.81
Zanand K 12.5	19/07/2011	2.28	0.99
	12/09/2011	0.73	0.46
	13/09/2011	1.00	0.55
	28/09/2011	1.40	0.79
	13/09/2012	1.86	0.95
	27/09/2012	1.69	0.75

Annex E. Night storage system

Table E.1. Deposition during the first hour

Distance (m)	Bed width (m)	Discharge (m³/s)	Water depth (m)	shear velocity (m/s)	Shear stress (N/m²)	Area (m²)	Sediment load (kg/s)	Concentration (kg/m³)	Deposition (m³/hr)	Deposition per 46 m (m³)	Deposition per reach (m³)
						Reach-1					
0	2	0.460	1.28	0.0026	0.0069	5.81	2.76	6.00	0.03	0.31	1.20
400	2	0.427	1.34	0.0016	0.0024	6.37	2.75	6.44	0.04	0.39	
800	2	0.393	1.43	0.0012	0.0015	6.97	2.74	6.96	0.05	0.49	
1200	2	0.362	1.51	0.0010	0.0010	7.56	2.72	7.51	0.06		
						Reach-2					
1200	2	0.362	1.25	0.0013	0.0017	5.60	2.72	7.51	0.05	0.83	5.47
1800	2	0.288	1.37	0.0009	0.0008	6.46	2.70	9.39	0.08	1.50	
2400	2	0.216	1.49	0.0006	0.0004	7.38	2.68	12.39	0.15	3.15	
3000	2	0.146	1.60	0.0004	0.0002	8.32	2.63	17.99	0.33		
						Reach-3					
3000	1.5	0.146	1.14	0.0007	0.0005	4.28	2.63	17.99	0.17	1.53	3.65
3500	1.5	0.188	1.19	0.0014	0.0020	4.59	2.57	13.67	0.11	1.05	
4000	1.5	0.229	1.23	0.0019	0.0038	4.89	2.53	11.06	0.08	1.06	
4700	1.5	0.285	1.30	0.0022	0.0046	5.32	2.51	8.79	0.06		
						Reach-4					
4700	1	0.285	1.24	0.0028	0.0080	4.33	2.51	8.79	0.05	0.47	1.86
5100	1	0.270	1.34	0.0016	0.0026	4.96	2.49	9.22	0.06	0.56	
5500	1	0.264	1.44	0.0003	0.0001	5.59	2.47	9.36	0.07	0.84	
6000	1	0.251	1.56	0.0010	0.0010	6.43	2.45	9.75	0.09		

Impact of improved operation and maintenance on cohesive sediment transport

Table E.2. Deposition during the second hour

Distance (m)	Bed width (m)	Discharge (m³/s)	Water depth (m)	shear velocity (m/s)	Shear stress (N/m²)	Area (m²)	Sediment load (kg/s)	Concentration (kg/m³)	Deposition (m³/hr)	Deposition per 46 m (m³)	Deposition per reach (m³)
						Reach-1					
0	2	0.46	1.30	0.0029	0.0085	5.98	2.76	6.00	0.03	0.31	1.02
400	2	0.453	1.38	0.0026	0.0069	6.55	2.75	6.07	0.04	0.34	
800	2	0.446	1.46	0.0024	0.0057	7.15	2.74	6.14	0.04	0.38	
1200	2	0.437	1.53	0.0022	0.0047	7.75	2.72	6.23	0.05		
						Reach-2					
1200	2	0.437	1.30	0.0028	0.0078	5.98	2.72	6.23	0.04	0.56	2.51
1800	2	0.388	1.42	0.0022	0.0048	6.87	2.71	6.99	0.05	0.80	
2400	2	0.339	1.54	0.0017	0.0029	7.82	2.69	7.95	0.07	1.16	
3000	2	0.29	1.66	0.0013	0.0017	8.83	2.67	9.21	0.11		
						Reach-3					
3000	1.5	0.29	1.12	0.0027	0.0072	4.21	2.67	9.21	0.05	0.58	2.32
3500	1.5	0.279	1.17	0.0024	0.0058	4.51	2.65	9.51	0.06	0.66	
4000	1.5	0.268	1.22	0.0022	0.0048	4.81	2.63	9.83	0.06	1.08	
4700	1.5	0.253	1.28	0.0019	0.0037	5.22	2.61	10.33	0.08		
						Reach-4					
4700	1	0.253	1.24	0.0023	0.0053	4.30	2.61	10.33	0.06	0.59	2.16
5100	1	0.253	1.34	0.0020	0.0041	4.92	2.59	10.24	0.07	0.66	
5500	1	0.253	1.44	0.0018	0.0032	5.58	2.57	10.15	0.08	0.92	
6000	1	0.253	1.56	0.0016	0.0024	6.43	2.54	10.05	0.09		

Table E.3. Deposition during the third hour

Distance (m)	Bed width (m)	Discharge (m³/s)	Water depth (m)	shear velocity (m/s)	Shear stress (N/m²)	Area (m²)	Sediment load (kg/s)	Concentration (kg/m³)	Deposition (m³/hr)	Deposition per 46 m (m³)	Deposition per reach (m³)
						Reach-1					
0	2	0.46	1.30	0.0029	0.0085	5.98	2.76	6.00	0.03	0.30	1.02
400	2	0.459	1.38	0.0027	0.0070	6.59	2.75	5.99	0.04	0.34	
800	2	0.45	1.46	0.0024	0.0057	7.18	2.74	6.08	0.04	0.38	
1200	2	0.431	1.54	0.0021	0.0045	7.78	2.72	6.32	0.05		
						Reach-2					
1200	2	0.431	1.33	0.0027	0.0071	6.20	2.72	6.32	0.04	0.55	2.11
1800	2	0.408	1.45	0.0022	0.0049	7.12	2.71	6.64	0.05	0.69	
2400	2	0.385	1.57	0.0019	0.0034	8.07	2.69	7.00	0.06	0.86	
3000	2	0.361	1.69	0.0016	0.0025	9.06	2.68	7.41	0.07		
						Reach-3					
3000	1.5	0.361	1.150	0.0032	0.0099	4.36	2.68	7.41	0.04	0.42	1.77
3500	1.5	0.338	1.20	0.0028	0.0078	4.66	2.66	7.88	0.04	0.52	
4000	1.5	0.307	1.24	0.0024	0.0058	4.95	2.65	8.63	0.05	0.83	
4700	1.5	0.306	1.31	0.0022	0.0049	5.37	2.63	8.60	0.06		
						Reach-4					
4700	1	0.306	1.23	0.0027	0.0075	4.28	2.63	8.60	0.05	0.42	1.57
5100	1	0.306	1.34	0.0024	0.0057	4.93	2.62	8.55	0.05	0.48	
5500	1	0.306	1.44	0.0021	0.0045	5.59	2.60	8.50	0.06	0.67	
6000	1	0.305	1.56	0.0018	0.0034	6.43	2.58	8.46	0.07		

Table E.4. Deposition during the fourth hour

Distance (m)	Bed width (m)	Discharge (m³/s)	Water depth (m)	shear velocity (m/s)	Shear stress (N/m²)	Area (m²)	Sediment load (kg/s)	Concentration (kg/m³)	Deposition (m³/hr)	Deposition per 46 m (m³)	Deposition per reach (m³)
Reach-1											
0	2	0.46	1.31	0.0029	0.0084	6.02	2.76	6.00	0.03	0.30	0.98
400	2	0.46	1.38	0.0027	0.0070	6.60	2.75	5.98	0.04	0.33	
800	2	0.46	1.46	0.0024	0.0060	7.18	2.74	5.95	0.04	0.35	
1200	2	0.46	1.54	0.0023	0.0051	7.79	2.72	5.92	0.04		
Reach-2											
1200	2	0.46	1.35	0.0028	0.0077	6.32	2.72	5.92	0.03	0.49	1.72
1800	2	0.449	1.47	0.0024	0.0057	7.22	2.71	6.04	0.04	0.57	
2400	2	0.439	1.58	0.0021	0.0043	8.15	2.70	6.15	0.05	0.66	
3000	2	0.43	1.70	0.0018	0.0033	9.18	2.68	6.24	0.05		
Reach-3											
3000	1.5	0.43	1.18	0.0036	0.0127	4.53	2.68	6.24	0.03	0.32	1.29
3500	1.5	0.409	1.22	0.0032	0.0104	4.81	2.67	6.54	0.03	0.36	
4000	1.5	0.389	1.27	0.0029	0.0083	5.13	2.66	6.85	0.04	0.61	
4700	1.5	0.362	1.33	0.0025	0.0063	5.53	2.65	7.33	0.04		
Reach-4											
4700	1	0.362	1.24	0.0032	0.0101	4.32	2.65	7.33	0.03	0.32	1.19
5100	1	0.36	1.34	0.0028	0.0077	4.93	2.64	7.33	0.04	0.36	
5500	1	0.36	1.44	0.0025	0.0061	5.59	2.63	7.30	0.04	0.51	
6000	1	0.36	1.56	0.0022	0.0046	6.43	2.61	7.26	0.05		

Table E.5. Deposition during the sixth hour

Distance (m)	Bed width (m)	Discharge (m³/s)	Water depth (m)	shear velocity (m/s)	Shear stress (N/m²)	Area (m²)	Sediment load (kg/s)	Concentration (kg/m³)	Deposition (m³/hr)	Deposition per 46 m (m³)	Deposition per reach (m³)
						Reach-1					
0	2	0.46	1.30	0.0029	0.0085	5.98	2.76	6.00	0.03	0.30	0.98
400	2	0.46	1.38	0.0027	0.0071	6.57	2.75	5.98	0.04	0.33	
800	2	0.46	1.46	0.0024	0.0060	7.20	2.74	5.95	0.04	0.35	
1200	2	0.46	1.54	0.0023	0.0051	7.80	2.72	5.92	0.04		
						Reach-2					
1200	2	0.46	1.36	0.0027	0.0075	6.38	2.72	5.92	0.03	0.48	1.63
1800	2	0.46	1.47	0.0024	0.0059	7.26	2.71	5.90	0.04	0.54	
2400	2	0.46	1.59	0.0021	0.0046	8.24	2.70	5.87	0.04	0.61	
3000	2	0.455	1.71	0.0019	0.0036	9.27	2.68	5.90	0.05		
						Reach-3					
3000	1.5	0.455	1.21	0.0036	0.0129	4.74	2.68	5.90	0.03	0.29	1.11
3500	1.5	0.446	1.26	0.0033	0.0110	5.06	2.68	6.00	0.03	0.32	
4000	1.5	0.437	1.30	0.0031	0.0095	5.36	2.67	6.10	0.03	0.50	
4700	1.5	0.424	1.37	0.0028	0.0077	5.78	2.66	6.27	0.03		
						Reach-4					
4700	1	0.424	1.24	0.0037	0.0134	4.33	2.66	6.27	0.03	0.24	0.90
5100	1	0.424	1.34	0.0032	0.0104	4.93	2.65	6.25	0.03	0.27	
5500	1	0.424	1.44	0.0029	0.0082	5.59	2.64	6.22	0.03	0.38	
6000	1	0.424	1.56	0.0025	0.0062	6.43	2.63	6.20	0.04		

Impact of improved operation and maintenance on cohesive sediment transport

Table E.6. Deposition during the eighth hour

Distance (m)	Bed width (m)	Discharge (m³/s)	Water depth (m)	shear velocity (m/s)	Shear stress (N/m²)	Area (m²)	Sediment load (kg/s)	Concentration (kg/m³)	Deposition (m³/hr)	Deposition per 46 m (m³)	Deposition per reach (m³)
						Reach-1					
0	2	0.46	1.3	0.0029	0.0085	5.98	2.76	6.00	0.03	0.30	0.98
400	2	0.46	1.38	0.0027	0.0071	6.57	2.75	5.98	0.04	0.33	
800	2	0.46	1.464	0.0024	0.0059	7.21	2.74	5.95	0.04	0.35	
1200	2	0.46	1.537	0.0023	0.0051	7.80	2.72	5.92	0.04		
						Reach-2					
1200	2	0.46	1.356	0.0027	0.0075	6.39	2.72	5.92	0.03	0.48	1.63
1800	2	0.46	1.476	0.0024	0.0058	7.31	2.71	5.90	0.04	0.54	
2400	2	0.46	1.59	0.0021	0.0046	8.24	2.70	5.87	0.04	0.60	
3000	2	0.46	1.71	0.0019	0.0037	9.27	2.68	5.84	0.05		
						Reach-3					
3000	1.5	0.46	1.22	0.0036	0.0128	4.81	2.68	5.84	0.03	0.29	1.06
3500	1.5	0.456	1.272	0.0033	0.0111	5.14	2.68	5.87	0.03	0.31	
4000	1.5	0.452	1.318	0.0031	0.0098	5.45	2.67	5.90	0.03	0.47	
4700	1.5	0.448	1.38	0.0029	0.0083	5.88	2.66	5.93	0.03		
						Reach-4					
4700	1	0.448	1.243	0.0038	0.0148	4.33	2.66	5.93	0.02	0.22	0.82
5100	1	0.447	1.343	0.0034	0.0114	4.95	2.65	5.93	0.03	0.25	
5500	1	0.447	1.439	0.0030	0.0090	5.58	2.64	5.91	0.03	0.35	
6000	1	0.448	1.56	0.0026	0.0069	6.43	2.63	5.87	0.03		

Table E.7. Deposition during the 12th hour

Distance (m)	Bed width (m)	Discharge (m³/s)	Water depth (m)	shear velocity (m/s)	Shear stress (N/m²)	Area (m²)	Sediment load (kg/s)	Concentration (kg/m³)	Deposition (m³/hr)	Deposition per 46 m (m³)	Deposition per reach (m³)
						Reach-1					
0	2	0.46	1.30	0.0029	0.0085	5.98	2.76	6.00	0.03	0.30	0.98
400	2	0.46	1.38	0.0027	0.0071	6.57	2.75	5.98	0.04	0.33	
800	2	0.46	1.46	0.0024	0.0060	7.20	2.74	5.95	0.04	0.35	
1200	2	0.46	1.54	0.0023	0.0051	7.80	2.72	5.92	0.04		
						Reach-2					
1200	2	0.46	1.36	0.0027	0.0075	6.39	2.72	5.92	0.03	0.48	1.63
1800	2	0.46	1.48	0.0024	0.0058	7.30	2.71	5.90	0.04	0.54	
2400	2	0.46	1.59	0.0021	0.0046	8.24	2.70	5.87	0.04	0.60	
3000	2	0.46	1.71	0.0019	0.0037	9.27	2.68	5.84	0.05		
						Reach-3					
3000	1.5	0.46	1.23	0.0035	0.0124	4.89	2.68	5.84	0.03	0.29	1.04
3500	1.5	0.46	1.28	0.0033	0.0110	5.20	2.68	5.82	0.03	0.30	
4000	1.5	0.46	1.33	0.0031	0.0099	5.51	2.67	5.80	0.03	0.45	
4700	1.5	0.46	1.39	0.0029	0.0086	5.93	2.66	5.78	0.03		
						Reach-4					
4700	1	0.46	1.24	0.0039	0.0155	4.33	2.66	5.78	0.02	0.21	0.78
5100	1	0.46	1.34	0.0035	0.0120	4.95	2.65	5.76	0.03	0.24	
5500	1	0.46	1.44	0.0031	0.0095	5.59	2.64	5.74	0.03	0.33	
6000	1	0.46	1.56	0.0027	0.0072	6.43	2.63	5.72	0.03		

Annex F.   Samenvatting

Efficiënt beheer en onderhoud van irrigatie systemen is nodig voor het verbeteren van het waterloopkundig functioneren van de kanalen, het verbeteren van de opbrengst van de gewassen en het verzekeren van een duurzame productie. Er is een grote behoefte om het onderzoek op een hoger peil te brengen en aan een verscheidenheid aan instrumenten, zoals apparatuur voor het regelen van de waterstromen, beslissingsondersteunende systemen, evenals aan veldonderzoek en beoordeling technieken. Waterbeheer wordt moeilijk als het gaat om sediment transport in irrigatiekanalen. De meeste studies simuleren het sediment transport van relatief grove korrelgroottes. Het sediment probleem in irrigatiekanalen wordt ingewikkelder als het gaat om transport van cohesief sediment. Daarom is meer onderzoek nodig voor een beter begrip van het gedrag van cohesief sediment transport onder een verscheidenheid aan beheer condities.

Dit onderzoek is uitgevoerd in het Gezira irrigatie systeem in Soedan. Het systeem, dat een van de grootste irrigatie systemen onder één beheerder in de wereld is, ligt in de droge en semi-droge regio. Het systeem is gekozen als studie object, omdat het kan fungeren als model voor soortgelijke irrigatie systemen. Het systeem heeft een totale oppervlakte van 880.000 ha en gebruikt 35% van de huidige toedeling aan Soedan van het water van de Nijl. Dit vertegenwoordigt 6-7 miljard m^3 per jaar. Het systeem wordt geïrrigeerd vanuit de Blauwe Nijl, die wordt gekenmerkt door een hoog gehalte aan fijn sediment. Het systeem wordt geconfronteerd met ernstige accumulatie van sediment in de irrigatiekanalen, wat een uitdaging vormt voor degenen die verantwoordelijk zijn voor het beheer en onderhoud van de kanalen. Elk jaar zijn grote investeringen nodig om het kanaal systeem te onderhouden en te verbeteren, teneinde het in een aanvaardbare conditie te houden. Een grote hoeveelheid cohesief sediment komt elk jaar in het systeem. Volgens eerdere studies wordt ongeveer 60% van het sediment afgezet in de irrigatiekanalen. De sediment accumulatie in de kanalen vermindert de transport capaciteit, veroorzaakt problemen voor de irrigatie, creëert ongelijkheid en onvoldoende watervoorziening en verhoogt de snelheid van aquatische onkruidgroei. De sedimentatie problemen hebben niet alleen ernstige gevolgen voor het functioneren van de irrigatiekanalen, maar brengen ook de duurzaamheid in gevaar, alsmede de gewasproductie. Twee kanalen in het systeem zijn geselecteerd om in detail te worden bestudeerd; het Zananda zijkanaal dat water afneemt van het Gezira hoofdkanaal, 57 km van de inlaat in de Sennar dam en het Toman tertiare kanaal op 12.5 km van de inlaat van hat Zananda zijkanaal.

De hypothese van de studie is dat het beheer en onderhoud van een irrigatie systeem een grote invloed heeft op het hydrodynamische gedrag van de kanalen en dus op de sediment verplaatsing en depositie. Het doel van deze studie was om de procedures voor beheer en onderhoud the verbeteren voor een verbeterd beheer van sediment en water. Dit kan worden bereikt door een beter begrip van de sedimentatie processen in de irrigatiekanalen van het Gezira irrigatie systeem en een duidelijk inzicht in het verband tussen het beheer van het irrigatiesysteem en het resulterende functioneren in termen van transport van het cohesieve sediment.

Het verzamelen van gegevens en de veldmetingen zijn uitgevoerd gedurende het vloed seizoen tussen juni en oktober in 2011 en 2012. Sediment bemonstering en metingen van het waterpeil zijn dagelijks op geselecteerde locaties uitgevoerd. De handmatig geregistreerde waterstanden omvatten ongeveer 1080 metingen per jaar. Daarnaast zijn ongeveer 1290 sedimentmonsters op verschillende locaties tijdens de studie periode geanalyseerd. Metingen in dwarsdoorsneden zijn aan het begin en einde van het vloed seizoen uitgevoerd om de variatie in ruimte en tijd van de sedimentatie in de bestudeerde kanalen te benaderen en veranderingen in het bodem profiel te bepalen. Het hoofd verdeelwerk en de aflaatwerken zijn gekalibreerd met behulp van de gemeten niveau afvoer relaties. De cohesieve eigenschappen van het sediment zijn nader uitgewerkt en de dominante factoren die afzetting in de irrigatiekanalen veroorzaken zijn geïdentificeerd. Sediment eigenschappen zoals korrelgrootteverdeling, mechanische en fysisch-chemische eigenschappen van het sediment zijn getest. De toediening van irrigatie water, het gecultiveerde gebied en de zaaidata voor de verschillende gewassen zijn verzameld. Andere gegevens zoals ontwerpgegevens van de kanalen, historische gegevens van het sediment en de stroming door bepaalde kanalen zijn geanalyseerd.

Uit de analyse van de gegevens blijkt een variatie in het waterniveau in de bestudeerde kanalen. Hierbij is van belang dat het beheer van het Gezira irrigatie systeem is gebaseerd op bovenstroomse regeling van de wateraanvoer. Uit de veldgegevens blijkt dat de watertoevoer naar het systeem niet regelmatig op een systematisch wijze bijgesteld wordt om aan de vraag te voldoen en het gewenste waterpeil te onderhouden. Continue verandering in de instelling van de schuiven resulteert in instabiliteit van het waterpeil. Deze situatie werd nog verergerd door meer sediment afzetting. Het waterpeil is tot ver boven het ontwerp niveau verhoogd en er zijn vooral bij de zijkanalen en de tertiaire kanalen afwijkingen in de niveaus voor het beheer. De stijging blijkt aan het begin van de bestudeerde zijkanalen en de tertiaire kanalen te variëren tussen 1,2 en 1,6 m boven het ontwerpniveau. Bovendien is vermindering van de waterdiepte in de kanalen aangetroffen als gevolg van het omhoog komen van de bodem en vergroting van het kanaal profiel door verkeerde verwijdering van het sediment. De resultaten tonen aan dat de wateraanvoer tijdens het vloed seizoen van 2011, vooral tijdens de periode waarbij een hoge sediment concentratie optrad, zeer groot was ten opzichte van de werkelijke waterbehoefte van de gewassen. De verhouding in de wateraanvoer wees in 2011 tijdens het grootste deel van de tijd op een overaanbod in het zijkanaal. De studie heeft ook een aantal waardevolle inzichten in de aard van het sediment in het Gezira irrigatie systeem opgeleverd.

Er is een beperking in de bestaande modellen die betrekking hebben op fijn sediment in irrigatiekanalen. De meeste sediment transport modellen zijn ontwikkeld voor estuaria en rivieren. Er was daarom grote behoefte aan een eenvoudig maar effectief numeriek model met daarin de waterbouwkundige kunstwerken om het transport van fijn sediment in irrigatiekanalen simuleren. Hoewel er overeenkomsten zijn tussen rivieren en irrigatiekanalen, zijn irrigatiekanalen toch anders. De aanwezigheid van een groot aantal waterbouwkundige kunstwerken om de waterstroming te controleren en de grote invloed van de zijkanten van de kanalen op de snelheidsverdeling in de waterstroming creëren enkele verschillen in beide soorten waterlopen. Daarom was het belangrijk om een model te ontwikkelen voor het simuleren van fijn sediment in irrigatiekanalen, met daarin verschillende soorten waterbouwkundige kunstwerken.

Op basis hiervan is het eendimensionale numerieke model voor het transport van fijn sediment (FSEDT) voor het omgaan met fijn sediment transport in irrigatiekanalen ontwikkeld. Het model is gebruikt als een instrument om het proces van water en sedimentafzetting onder verschillende beheer en onderhoud scenario's te simuleren. Het profiel van het oppervlakte water is voorspeld met de predictor-corrector methode om de vergelijking voor geleidelijk variërende stroming op te lossen. De slibconcentratie is voorspeld op basis van de oplossing van de eendimensionale advectie-diffusie vergelijking. De uitwisseling van bodemmateriaal is gebaseerd op de Partheniades (1962) en Krone (1965) vergelijkingen bepaald. De verandering in het niveau van de bodem is op basis van de vergelijking voor de massabalans voor het sediment berekend, die numeriek is opgelost door gebruik te maken van de eindige verschillen methode. Het model is toegepast op het Gezira irrigatie systeem. Op basis van veldgegevens is het model is gekalibreerd en gevalideerd. Het voorspelde bodem profielen vertoonden een goede overeenkomst met de gemeten profielen. Het model is in staat om het bodem profiel voor iedere periode van simulatie te voorspellen. Het model kan de curve voor de sedimentconcentratie op verschillende punten in een kanaalpand voorspellen, naast het totale volume van de sediment afzetting in het pand. De uitvoer van het model kan worden gepresenteerd in tabellen of grafieken.

Het sediment transport in de irrigatiekanalen is uitgaande van verschillende scenario's gesimuleerd. De onderlinge relatie tussen de waterstroming en het sediment transport is onder veranderende stromingsomstandigheden in de irrigatiekanalen onderzocht. Twee beheer scenario's voor het zijkanaal zijn in de studie getest. Met het model is het verspringende systeem voor de wateraanvoer dat vele jaren in het Gezira irrigatie systeem in verband met sediment afzetting is toegepast geëvalueerd. Een ander voorgesteld scenario op basis van de waterbehoefte van de gewassen is ook getest. Daarnaast, is het beheer onder in de toekomst gewijzigde omstandigheden door verlaging van de sediment concentratie getest. De verschillende beheer scenario's zijn in termen van sedimentatie vergeleken met de bestaande toestand op basis van de tijdens het vloed seizoen in 2011 verzamelde gegevens. Op basis hiervan worden de volgende opmerkingen gemaakt:

- het effect van verschillende instellingen van de bovenkant van de beweegbare stuwen is onderzocht en geringere sediment afzetting is gevonden wanneer de bovenkant van de stuwen op de laagste stand werd ingesteld. Het sediment transport in de kanalen wordt, vooral bovenstrooms van de beweegbare stuwen, beïnvloed door het beheer van de waterbouwkundige kunstwerken. Het effect zet zich tot ongeveer 3 km bovenstrooms van de stuw voort;
- gedurende vele jaren is, gebaseerd op de behoefte en het gecultiveerde gebied, het verspringende systeem van toediening van water in het Gezira irrigatie systeem toegepast. Dit systeem van beheer is echter in de afgelopen jaren niet meer toegepast. De simulatie van het transport van het opgeloste sediment toonde aan dat minder sediment wordt afgezet wanneer dit systeem wordt toegepast. De afname van de sedimentatie bleek in vergelijking met de werkelijke situatie in 2011 respectievelijk 34 en 40% in het eerste en tweede pand te zijn. Toen het model werd ingesteld op de kanaal profielen uit het ontwerp trad meer sedimentatie op. De helling van de bodem werd 13 cm/km en 18 cm/km voor respectievelijk het eerste en tweede pand terwijl dit in het ontwerp 10 cm/km en 5 cm/km was. De steilere helling in de bodem, speciaal die in het tweede pand is een indicator voor onzorgvuldig herprofileren;
- de reductie van de wateraanvoer gedurende de periode van hoge concentratie tussen 10 juli en 10 augustus, op basis van de waterbehoefte van het gewas in vergelijking met de situatie in 2011 resulteert in een afname van de sedimentatie met respectievelijk 51 en 55% voor het eerste en tweede pand;
- de vermindering van de sediment concentratie in de Blauwe Nijl met 50% ten gevolge van de bouw van de Renaissance Dam in Ethiopië en/of verbetering van het grondgebruik is gesimuleerd. De resultaten van de simulatie van het sediment transport in het zijkanaal geven aan dat de sedimentatie in respectievelijk het eerste en tweede pand in vergelijking met de situatie in 2011 74 en 81% lager is.

In de tertiaire kanalen, zijn de stuwen voor de nacht opslag dwars in deze kanalen ontworpen. Het idee achter het nacht opslagsysteem was om door het om 18:00 uur sluiten van alle uitlaat pijpen naar de velden en de schuiven in de tertiaire kanalen om water in de nacht op te slaan en deze weer te openen om 6:00 uur. Hoewel dit systeem is verdwenen om ruimte te bieden voor intensivering van de verbouw van gewassen en om het hoofd te bieden aan de verslechtering van de wateraanvoer ten gevolge van het slechte onderhoud van de kanalen, is dit scenario ook gesimuleerd. De hydrodynamische stroming in de kanalen gedurende de vultijd is gesimuleerd met het DUFLOW model, omdat dit model kan worden toegepast op niet stationaire stroming. Er is een spreadsheet ontworpen om de afzetting per uur te voorspellen op basis van de resultaten van het DUFLOW model. Voor wat betreft het sediment transport is het nacht opslagsysteem naast andere scenario's vergeleken met het continue systeem. Er is gevonden dat:

- in vergelijking met het nacht opslagsysteem vermindert het continue systeem de hoeveelheid afgezet sediment met 55%;
- toen het systeem werd afgestemd op de waterbehoefte van het gewas in 2011 nam de sedimentatie met ongeveer 29% af;
- de sedimentatie in geringe mate toenam bij verminderde capaciteit van de uitlaat pijpen naar de velden. De toename van sedimentatie is respectievelijk 22% en 31% wanneer de capaciteit van de uitlaat pijpen naar de velden wordt verminderd met 75% en 50%.

De belangrijkste bevindingen en de bijdragen die door deze studie worden geleverd zijn:

- de studie resulteert in een model voor het beschrijven van cohesief sediment voor een effectief beheer van sediment en water in irrigatiekanalen, dat voor het omgaan met fijn sediment op grote schaal kan worden toegepast bij soortgelijke irrigatieprojecten;
- het is mogelijk om het beheer van sediment en water te verbeteren door het verbeteren van het beheer en onderhoud. De hoge irrigatie-efficiëntie vertoont de neiging om de sediment belasting door de wateraanvoer te beperken en ten gevolge daarvan wordt minder sedimentatie verwacht;
- de studie resulteerde in strategieën voor het waterbeheer dat de sedimentatie in irrigatiekanalen kan verminderen door het toepassen van het continue irrigatie systeem op basis van de waterbehoefte van de gewassen gedurende de periode met hoge sediment concentratie waarbij de uitlaat pijpen naar de velden op hun volle capaciteit werken.

Het ontbreken van activiteiten voor goed onderhoud en waterbeheer speelt een prominente rol bij het verhogen van de sedimentatie in de irrigatiekanalen in het Gezira irrigatie systeem. Verbeteren van het beheer en onderhoud is niet alleen de manier om de sedimentatie in de irrigatiekanalen te beperken. Er zal veel aandacht moeten worden besteed aan het verbeteren van het ontwerp, omdat de voorwaarden die de basis vormden voor het oorspronkelijke ontwerp met de tijd zijn veranderd, zoals de werking van het systeem (nacht opslagsysteem, verspringend system), intensiteit van het verbouwen van gewassen en de geometrie van de kanalen. Met andere woorden, herstel van het systeem zal niet één van de oplossingen bieden om de accumulatie van de sedimentatie in de kanalen het hoofd te bieden, maar het systeem zelf dient opnieuw te worden vorm gegeven. Het ontwikkelde model kan worden gebruikt om het nieuwe ontwerp te beoordelen en de voorgestelde beheersplannen te waarderen qua transport van het cohesieve sediment.

Annex G.   About the author

Ishraga Sir Elkhatim Osman was born in Port Sudan, Sudan in 1979. She received her Bachelor degree from the University of Khartoum, Civil Engineering Department in 2002. In June 2007 she was awarded a master degree in Water Resources Engineering from the University of Khartoum. Ishraga worked as a researcher at the UNESCO Chair in Water Resources in Sudan from 2003 to 2009. She has engaged in a number of projects and became a member of the River Morphology Cluster of the Nile Basin Capacity Building Network. She was involved in the "Nile Basin Reservoir Sedimentation Prediction and Mitigation Project" in addition to the local action project. She participated in many activities in the FRIEND/NILE Project particularly within its sediment transport and watershed management component. She was also involved in several researches related to water management in irrigation canals and assessment of irrigation performance. She is a lecturer at the Civil Engineering Department, University of Khartoum since 2009 where she is engaged in teaching, research, consultancy and supervision of BSc theses of graduating students.

In May 2010, she started her PhD study in the Hydraulic Engineering - Land and Water Development Core of the Water Science and Engineering Department, UNESCO-IHE Institute for Water Education in Delft, the Netherlands. During her PhD study, she attended and gave oral presentations in two international conferences. She also submitted two articles related to her PhD research to international peer-reviewed journals and she is currently preparing two more journal papers. Ishraga has attended several courses within the SENSE network.

Netherlands Research School for the
Socio-Economic and Natural Sciences of the Environment

D I P L O M A

For specialised PhD training

The Netherlands Research School for the
Socio-Economic and Natural Sciences of the Environment
(SENSE) declares that

Ishraga Sir ElKhatim Osman

born on 1 January 1979 in Port Sudan, Sudan

has successfully fulfilled all requirements of the
Educational Programme of SENSE.

Delft, 3 September 2015

the Chairman of the SENSE board

Prof. dr. Huub Rijnaarts

the SENSE Director of Education

Dr. Ad van Dommelen

The SENSE Research School has been accredited by the Royal Netherlands Academy of Arts and Sciences (KNAW)

KONINKLIJKE NEDERLANDSE
AKADEMIE VAN WETENSCHAPPEN

The SENSE Research School declares that Mr Ishraga Osman has successfully fulfilled all requirements of the Educational PhD Programme of SENSE with a work load of 41.6 EC, including the following activities:

<u>SENSE PhD Courses</u>

o   Environmental Research in Context (2013)
o   Research in Context Activity: Co-organising communication and media contacts for International conference on: New Nile Perspectives - Scientific advances in the Eastern Nile Basin, Sudan (2013)
o   SENSE writing week (2014)

<u>Other PhD and Advanced MSc Courses</u>

o   Capacity building course: Research formulation, methods and skills, University of Khartoum and University of Reading (2010)
o   Conveyance systems -Sediment transport in irrigation canal, UNESCO-IHE, Delft (2010)
o   Irrigation and drainage structures, UNESCO-IHE, Delft (2010)
o   Service oriented management of irrigation systems, UNESCO-IHE, Delft (2011)
o   Morphological modelling using Delft 3D model, UNESCO-IHE, Delft (2013)

<u>Management and Didactic Skills Training</u>

o   Supervising three BSc students on a project entitled 'Estimating of the crop water requirement in Gezira Irrigation Scheme', Alzaiem Alazhari University (2013-2014)
o   Teaching in BSc course 'Hydraulic-II', University of Khartoum, Sudan (2011)
o   Guest lecture 'Field measurements and data collection techniques in MSc course 'Hydrometry', University of Khartoum (2015)

<u>Oral Presentations</u>

o   *The effect of Blue Nile sediment load in irrigation canals. Case study: Gezira Scheme.* Scientific Symposium on 'Eastern Nile Cooperation: Opportunities for Regional Development', 10-12 March 2014, Khartoum, Sudan
o   *Sedimentation in irrigation in Gezira Scheme, Sudan.* Scientific Meeting on The New Nile Perspectives: scientific advances in the Eastern Nile Basin, 6-7 May 2013, Khartoum, Sudan
o   *Improving the operation and maintenance for better sediment and water management in Gezira Scheme, Sudan.* 21st International Congress on Irrigation and Drainage (ICID), 15-23 October 2011, Tehran, Iran

SENSE Coordinator PhD Education

Dr. ing. Monique Gulickx

T - #0414 - 101024 - C200 - 240/170/11 - PB - 9781138028807 - Gloss Lamination